Praise for

Alive

"A great book. . . . [A]n incredible saga. Read's accomplishment in recording a struggle both physical and spiritual is superb." —*Philadelphia Inquirer*

"Terrifying . . . exciting and hugely entertaining." —*New York Times*

"More than an adventure story. . . . An intimate, unglorified account of the fear, tensions, and practical problems of people who want to live. . . . Not to be missed." —*Denver Post*

"Flawless. . . . Superb. . . . A testimonial to the durability and determination of young men who might have chosen to die and simply would not." —*Detroit Free Press*

"Tragic, heartwarming, inspiring, shocking, terrifying, and true." —*Columbus Dispatch*

"No one will come away unmoved by the book, and no one will be able to put it down. . . . There is no way of reading *Alive* without a heightened sense of one's own life and its value." —*The New Republic*

"An adventure story to rival those of Hillary and Tensing, Thor Heyerdahl, or Jacques Cousteau." —*Chicago Tribune*

"A chilling tale . . . a book of real and lasting value. . . . [It] will involve the reader as thoroughly as the best adventure novel."
—*Rolling Stone*

"A book you won't soon forget."
—*Cleveland Press*

"An adventure both physical and spiritual. . . . Mr. Read was chosen by the survivors to write a full and balanced account of their ordeal, and he has done so with conspicuous success."
—*The New Yorker*

"An extraordinary book. . . . Read writes in a sharply honed documentary style that is sometimes intensely dramatic, sometimes all the more effective for being deliberately underplayed."
—*Publishers Weekly*

"Excellent. . . . [It] goes beyond the sensationalism of the headlines to uncover the human truths of an excruciating experience."
—*Worcester Sunday Telegram*

"Quite a book. . . . [Read's] account of this ordeal is inspirational."
—*Playboy*

"As provocative as it is engrossing. . . . One of the classic epics of how men interact in the face of near certain death."
—*The Village Voice*

About the Author

PIERS PAUL READ is the author of a number of critically acclaimed works of fiction and nonfiction, most recently *The Templars*, a history of the crusading order; a novel, *Alice in Exile;* and *Alec Guinness: The Authorized Biography*. He lives in London.

ALIVE

Sixteen Men, Seventy-two Days, and Insurmountable Odds— the Classic Adventure of Survival in the Andes

PIERS PAUL READ

HARPER
PERENNIAL

HARPER PERENNIAL

Photographs numbers 1, 2, 3, and 4 were taken by Roberto Caldeyro Stajano; numbers 8, 9, and 10 were found in Dr. Francisco Nicola's camera. Numbers 28, 29, 31, 32, and 33 were taken by Parrado and Vizintin and are reproduced by permission of Sygma, Paris; numbers 35, 38, and 41 were taken by Jean-Pierre Laffont and are also reproduced by permission of Sygma, Paris. Numbers 28, 29, 32, and 33 are copyright by Gamma, Paris; number 41 is copyright by J.P. Laffont Gamma. Anthony Peagam took the photographs of the survivors numbered 12, 13, 15, 16, 19, 20, 21, 22, 23, 24, and 27, which are reproduced by permission of *The Observer* magazine, London. Dr. Gustavo Nicolich took the photographs numbered 37 and 42; number 40 is from Europa Press News Service.

First Harper Perennial edition published 2005.

Library of Congress Cataloging-in-Publication Data

Read, Piers Paul.
Alive/Piers Paul Read.
p. cm.
ISBN-10: 0-06-077866-0
ISBN-13: 978-0-06-077866-8
1. Survival after airplane accidents, shipwrecks, etc. 2. Aircraft accidents—Andes Region. 3. Cannibalism—Andes Region. I. Title.

TL553.9.R4 2005
982'.6—dc22 2005040230

08 09 RRD 10 9 8

We decided that this book should be written and the truth known because of the many rumors about what happened in the cordillera. We dedicate this story of our suffering and solidarity to those friends who died and to their parents who, at the time when we most needed it, received us with love and understanding.

Pedro Algorta, Roberto Canessa, Alfredo Delgado,
Daniel Fernández, Roberto Francois, Roy Harley,
José Luis Inciarte, Alvaro Mangino, Javier Methol,
Carlos Páez, Fernando Parrado, Ramón Sabella,
Adolfo Strauch, Eduardo Strauch,
Antonio Vizintín, Gustavo Zerbino

Montevideo, October 30, 1973

Acknowledgments

In writing this book I was helped by various people—especially by Edward Burlingame of J. B. Lippincott Company, who first suggested that I should write it.

In Montevideo my researches were aided by two Uruguayan journalists. The first was Antonio Mercader, whose assistance was suggested to me by the Committee of the Old Christians. He provided me not only with the complex details of the search for the plane undertaken by the parents but also with invaluable material on the background of the survivors. The second journalist was Eugenio Hintz, who gathered material about what was done by the official agencies of the Uruguayan and Chilean governments. I am also indebted to Rafael Ponce de León and Gerard Croiset, Jr., who told me of their roles in the search for the Fairchild; to Pablo Gelsi, who acted as my interpreter; and to Dr. Gilberto Regules, for his advice and friendship.

In London I was helped with the transcription of the tapes and the organization of the considerable material I had gathered in Uruguay by Georgiana Luke and later, in further research, by Kate Grimond.

I was given a free hand in writing this book by both the publisher and the sixteen survivors. At times I was tempted to fictionalize certain parts of the story because this might have added to their dramatic impact, but in the end I decided that the bare facts were sufficient to sustain the narrative. With the exception of the rendering of some speech in dialogue form, nothing in this book departs from the truth as it was told to me by those involved.

It is to them, finally, that I am most grateful. Everywhere I went in Uruguay I was met by "that intimate courtesy and native grace of manner" which W. H. Hudson encountered in the same country more than a hundred years ago. I found it in the families of those who died, in the families of the survivors, and above all in the survivors themselves, who treated me with a most exceptional warmth, candor, and trust.

When I returned in October 1973 to show them the manuscript of this book, some of them were disappointed by my presentation of their story. They felt that the faith and friendship which inspired them in the cordillera do not emerge from these pages. It was never my intention to underestimate these qualities, but perhaps it would be beyond the skill of any writer to express their own appreciation of what they lived through.

P. P. R.

Introduction

The story of how a group of young Uruguayans had emerged from the Andes after seventy days appeared in newspapers throughout the world shortly after Christmas 1972. I was then staying in my family home in the north of England—about as far from Montevideo as it is possible to be—and when it leaked out that the survivors had eaten the bodies of those killed in the accident, my immediate reaction was one of mild disgust. When my American publisher, Edward Burlingame, called from New York suggesting that there might be a book in their story, I dismissed the idea out of hand.

It was only thanks to Burlingame's persistence and the offer of a first class return ticket with no strings attached, that I flew to Montevideo in the middle of January to meet the survivors and let myself be put forward as a possible writer of an authorized account of their ordeal. I was immediately taken up by a round of conferences and social engagements: other more powerful American publishers—among them Harper & Row and *Reader's Digest*—were competing for the right to publish the survivors' story. I remember, in particular, a party given by the artist Carlos Páez Vilaró, the father of Carlitos, one of the survivors, in the extraordinary white house that he had designed himself, meeting the still-haggard survivors, their beautiful *novias,* and handsome healthy-looking friends, while drinking *pisco* sours and watching the sun set over the South Atlantic.

The survivors had appointed a committee to advise them on the different bids. Some of their parents were against a book of any kind, but since fanciful accounts of their ordeal were appearing in the newspapers, it was decided, as they say in their dedication to *Alive,* "that this book should be written and the truth known because of the many rumors about what had happened in the *cordillera.*"

The survivors' chief concern was that the account should not be "sensationalized" and due weight be given to the spiritual significance of their ordeal. On the face of it, a thirty-one-year-old English novelist who spoke no Spanish and had never written nonfiction was the least suitable among the writers put forward when some, such as Gay Talese, had established reputations in the field. However, some of the factors that might seem to have disqualified me, in fact, worked in my favor: I was an Englishman, not a *Yanqui;* I was closer to their age; and, above all, I shared with them the faith and practice of the Roman Catholic religion.

After ten days or so of frantic negotiations, I was chosen by the committee and an agreement was signed with the survivors. Edward Burlingame, the publisher who had so successfully presented me in the best possible light, also extracted a concession which some of the survivors later regretted—that I should not be expected to "violate my conscience" by agreeing to any form of censorship over the finished book.

At the beginning of February 1973, Burlingame returned to New York and I started my research, interviewing the survivors, their families, and also the families of those who had not returned. The process was difficult and at times painful: the survivors had been unwilling to talk to either priests, parents, or psychotherapists because to confess to the first would imply that they had sinned, to the second that they were still children and to the third that they were *loco*—insane! I now became their confidante, counselor, and confessor and, as a result, the bond forged with my subjects became so close that they sometimes called me "the seventeenth survivor."

When I returned to Montevideo, in October 1973 with Edward Burlingame and the typescript of *Alive,* some of them felt I had abused this trust. They were shocked that I included in my narrative what they called "the details"—the gruesome minutiae of their cuisine. They felt they had been portrayed as monsters. Where had I written that they were admirable and courageous young men? Why had I not described their survival as a miracle? And then, individually, each would take me aside to say that, while I had been accurate in depicting the characters of the other fifteen, I had wholly failed to understand *him*.

The atmosphere in our suite in the Hotel Victoria Plaza was chaotic, almost hysterical. We were told that if the book was published in its present form, the survivors would be stoned in the streets of Montevideo. Only Nando Parrado remained calm. To the others, I described the method I had adopted in writing the book: the story was so strong that it was best told unadorned. It was not for the author to tell the reader what to think, but for the reader to decide for himself. The "details" might be shocking, but if they were excluded, people might come to believe that worse things had happened on the mountain than those described in the book.

Alive was published in New York the following spring, and throughout the world soon after. Overwhelmingly, its reception was positive and the survivors, far from being stoned in the streets of Montevideo, became national heroes. Over the years it has sold over five million copies in fifteen different languages, and in 1992 was made into a film. Of the many letters I have received, not one has criticized the survivors for what they did. Quite to the contrary: as one reviewer put it, this story "of a group of human beings who rose, *in extremis,* to heights beyond their own, or anyone else's expectations . . . left me, at least, feeling a good deal better than usual about belonging to the human race."

Greater love hath no man than this,
that a man lay down his life for his friends.

—John 15:13

One

1

Uruguay, one of the smallest countries on the South American continent, was founded on the eastern bank of the River Plate as a buffer state between the emerging giants of Brazil and Argentina. Geographically it was a pleasant land, with cattle running wild over immense pasturelands, and its population lived modestly either as merchants, doctors, and lawyers in the city of Montevideo or as proud and restless gauchos on the range.

The history of the Uruguayans in the nineteenth century is filled first with fierce battles for their independence against Argentina and Brazil and then with equally savage civil skirmishes between the Blanco and Colorado parties, the Conservatives from the interior and the Liberals from Montevideo. In 1904 the last Blanco uprising was defeated by the Colorado president, José Batlle y Ordóñez, who then established a secular and democratic state which for many decades was regarded as the most advanced and enlightened in South America.

The economy of this welfare state depended upon the

pastoral and agricultural products which Uruguay exported to Europe, and while world prices for wool, beef, and hide remained high, Uruguay remained prosperous; but in the course of the 1950s the value of these commodities went down and Uruguay went into a decline. There was unemployment and inflation, which in turn gave rise to social discontent. The civil service was overstaffed and underpaid; lawyers, architects, and engineers—once the aristocracy of the nation—found themselves with little work and were paid too little for what there was. Many were compelled to choose secondary professions. Only those who owned land in the interior could be sure of their prosperity. The rest worked for what they could get in an atmosphere of economic stagnation and administrative corruption.

As a result, there arose the first and most notable movement of urban guerrilla revolutionaries, the Tupamaros, whose ambition was to bring down the oligarchy which governed Uruguay through the Blanco and Colorado parties. For a while things went their way. They kidnapped and ransomed officials and diplomats and infiltrated the police force, which was set against them. The government called upon the army, which ruthlessly uprooted these urban guerrillas from their middle-class homes. The movement was suppressed; the Tupamaros were locked away.

In the early 1950s a group of Catholic parents, alarmed at the atheistic tendencies of the teachers in the state schools—and dissatisfied with the teaching of English by the Jesuits—invited the Irish Province of the Christian Brothers to start a school in Montevideo. This invitation was accepted, and five Irish lay brothers came out from Ireland by way of Buenos Aires to found the Stella Maris College—a school for boys between the ages of nine and sixteen—in the suburb of Carrasco. In May of 1955 classes were started in a house on the *rambla* which looked out under vast skies over the South Atlantic.

Though they spoke only halting Spanish, these Irish Brothers were well suited to the task they now sought to perform. Uruguay might be far from Ireland, but it too was a small country with an agricultural economy. The Uruguayans ate beef as the Irish ate potatoes, and life here, like life in Ireland, was led at a gentle pace. Nor was the structure of that part of Uruguayan society to which they catered unfamiliar to the Brothers. The families who lived in the pleasant modern houses built amid the pine trees of Carrasco—the most desirable suburb of Montevideo—were mostly large, and there were strong bonds between parents and children which persisted through adolescence into maturity. The affection and respect which the boys felt for their parents was readily transferred to their teachers. This proved enough to maintain good behavior and, at the request of the parents of their pupils, the Christian Brothers gave up their long-standing use of the disciplinary cane.

It was also customary in Uruguay for young men and women to live with their parents even after they had left school, and it was not until they got married that they left home. The Christian Brothers often asked themselves how it was that, in a world where acrimony between generations sometimes seemed to be the spirit of the age, the citizens of Uruguay—or at least the residents of Carrasco—should be spared this conflict. It was as if the torrid vastness of Brazil to the north and the muddy waters of the River Plate to the south and west acted not only as natural barriers but as a protective shell in a cocoon of time.

Not even the Tupamaros troubled the Stella Maris College. The pupils, who came from Catholic families with conservative inclinations, had been sent by their parents to the Christian Brothers because of this order's traditional methods and old-fashioned objectives. Political idealism was more likely to flourish under the Jesuits, who trained the intellect, than under the Christian Brothers, whose aim was to build the character of their boys—and the generous use of corporal punishment, which they had abandoned at the request of the parents, was not the only means to this end at their disposal. The other was rugby football.

The game played at the Stella Maris College was and still is the same as that played in Europe. Two teams of fifteen men face one another on the field. They wear no helmets or protective padding, and there are no substitutes. The objective of each team is to place the oval ball on the try line defended by the other side or to kick the ball over the bar and between the posts of the H-shaped goal. The ball can be kicked, carried, or passed back; the player who holds it can be tackled by an opponent who will throw himself through the air to bring him down—grabbing him around the neck, the waist, or the legs. The only defense against a tackle is to dodge it or use the handoff, a hand pushed at the face or body of the tackling player.

If play stops—say, for instance, when the ball is passed forward—the whistle is blown by the referee and a scrum is formed. The forwards of the two teams lock together like a giant crab. In the front row, a hooker and two props dig their heads and shoulders into the heads and shoulders of their opposite numbers on the other side. Behind them comes the second line, their heads thrust between the buttocks of the front row. This human battering ram is held together by a "lock" at the back and buttressed by a wing forward on either side.

The scrum half of whichever team has the advantage throws the ball into the middle, between the two teams, whereupon the hooker kicks it back or the forwards push their rivals off the ball. The scrum half then retrieves it and throws it to the halfbacks, who run forward, passing the ball back along their line to the wing three-quarters to score a try.

It is a hard game—fine if played with skill, brutal if played clumsily. A broken leg or a broken nose is not uncommon; every scrum means hacked shins, every tackle leaves a player winded. It demands not only exceptional fitness to play at great speed for an hour and a half (including ten minutes' rest at halftime), but also self-control and team spirit. The man who places the ball on the touch line is not necessarily the finest player but more often just the last link in the chain.

When the Christian Brothers first came to Uruguay, rugby was hardly played at all there; indeed, they found themselves in a country where soccer was not just the national sport but a communal passion. Along with per capita consumption of beef, it was the only sphere in which Uruguay triumphed over the great nations of the world (they won the World Cup in 1930 and 1950), and to ask young Uruguayans to play a different game was like feeding them on bread and potatoes instead of their usual diet.

Having sacrificed one pillar of their educational system in giving up the cane, the Christian Brothers were not going to give up the other. They held to their contention that soccer was a sport for the prima donna, whereas rugby football would teach the boys to suffer in silence and work as a team. The parents expostulated but they acquiesced, and in time they even came to share the opinion of the Christian Brothers as to the merits of the game.

As for their sons, they played it with growing enthusiasm, and when the first generation had passed through the school, many of the graduates were unwilling to give up either rugby or the Stella Maris College. The idea of an old boys' group of alumni was conceived, and in 1965, ten years after the foundation of the school, this association came into being. It was called the Old Christians Club, and its chief activity was playing rugby on a Sunday afternoon.

As the years passed, these games became popular—even fashionable—and each summer brought new members to the Old Christians Club and a wider choice of players for a better team. Rugby itself caught on in Uruguay, and the Old Christians' first fifteen, with the shamrock on their shirts, became one of the best teams in the country. In 1968 they won the Uruguayan national championship, and again in 1970. Ambition grew with success. The team made a trip across the estuary of the River Plate to play teams in Argentina, and in 1971 they made up their minds to go farther afield and play in Chile. To make this possible and not too expensive, the club chartered a plane from the Uruguayan Air Force to fly them from Montevideo to Santiago, and tickets for seats not required by the team were sold to their friends and supporters. The trip was a great success. The team played the Chilean national team and the first fifteen of the Old Boys Grange, winning one match and losing the other. At the same time, they had a short holiday in a foreign land. For many it was their first journey abroad and their first sight of the snow-covered peaks and glacial valleys of the Andes. Indeed, the trip was such a success that no sooner had they returned to Montevideo than they planned to go again the following year.

2

By the end of the next season, considerable doubt surrounded their plans. The first fifteen of the Old Christians had, through overconfidence, lost the Uruguayan championship to a team they considered inferior; as a result, some of the club's officers thought that they did not deserve another trip to Chile. Another problem they faced was filling the forty-odd seats of the Fairchild F-227 which they had chartered from the Air Force. The cost of hiring the plane was U.S. $1,600. If forty seats were filled, it would only cost each passenger around $40 to fly to Santiago and back—less than a third of the commercial fare. The more seats that remained empty, the more it would cost each passenger, and they still had to meet the expenses of five days in Chile.

Word went around that the trip might have to be

canceled, whereupon those who wanted to go began to look for recruits among their friends, relations, and fellow students. There were various arguments for going to Chile. For the serious-minded students of economics there was the experiment in democratic Marxism under President Allende; for the less earnest there was the promise of high living at a low price. The Chilean escudo was weak: the dollar fetched a high price on the black market, and, as a sports delegation, the Old Christians would not be obliged to exchange their money at the official rate. The rugby players tempted their friends with visions of the pretty and uninhibited Chilean girls on the beaches of Viña del Mar or at the ski resort of Portillo. The net was cast wide, drawing in the mother and sister of one boy, the older cousins of another. By the day when the money had to be delivered to the Air Force, they had sold enough tickets to cover the cost.

At around six on the morning of Thursday, October 12, 1972, the passengers began to arrive in small groups at Carrasco airport for the second Old Christians' trip to Chile. They were driven in cars or pickup trucks by their parents and girlfriends, and their vehicles were parked beneath the palm trees outside the airport building, which, surrounded by large tracts of well-cut grass, looked more like the clubhouse of a golf course than an international airport. In spite of the early hour and the bleary looks on their faces, the boys were dressed smartly in slacks and sports coats, and they greeted one another with great spirit and excitement. The parents, too, all seemed to know one another. With fifty or sixty people talking and laughing together, it was almost as if someone had chosen the foyer of the airport to throw a party.

Calm amid all this confusion stood thc two somc-what stocky figures of Marcelo Pérez, the captain of the first fifteen, and Daniel Juan, the president of the Old Christians, who had come to see them off. Pérez looked decidedly happy. It was he who had been most enthusiastic about this trip to Chile and he who had suffered most at the prospect of its cancellation. Even now that it was taking place, the brow beneath his balding head would wrinkle as some problem was brought to his attention. One such problem was the absence of Gilberto Regules. The boy had not met his friends at the appointed time; he had not come to the airport; and now, when they telephoned his home, there was no reply.

Marcelo knew they could not wait for long. Their departure had to be early in the morning because it was dangcrous to fly through the Andes in the afternoon when the warm air of the Argentine plains rose to clash with the cold air of the mountains; already, the Fairchild had taxied across the tarmac from the military base which adjoined the civilian airport.

The boys milling around seemed a motley collection, ranging in age from eighteen to twenty-six, but they had more in common than met the eye. Most of them were Old Christians, and most of those who were not had been to the Jesuit College of the Sacred Heart in the center of Montevideo. Besides the team and its supporters, there were their friends, cousins of friends, and fellow students from the faculties of law, agriculture, economics, and architecture in which many of the Old Christians were now studying. Three of the boys were medical students, two of whom played on the team. Some of them had neighboring ranches in the interior; many more were neighbors in Carrasco, for what they all had in common was their class and their religion. They were, almost without exception, from the more prosperous section of the community, and all were Roman Catholic.

Not all the passengers who checked in at the desk of the Uruguayan Military Transport were Old Christians or even young men. There was a plump middle-aged woman, Señora Mariani, who had bought a ticket from the Air Force to go to her daughter's marriage to a political exile in Chile. There were two middle-aged couples and a tall pleasant-looking girl of around twenty named Susana Parrado, who stood in line with her mother, her brother Nando, and her father, who had come to see them off.

When their baggage had been checked in, the Parrados went up to the airport restaurant which overlooked the runway and ordered breakfast. At another table, a little distance from the Parrados, sat two students of economics who wore scruffier clothes than the rest, as if to show that they were socialists—a contrast to Susana Parrado, who wore a beautiful fur-lined coat made from antelope skin which she had bought only the day before.

Eugenia Parrado, her mother, had been born in the Ukraine, and both Susana and her brother were exceptionally tall, with fine, brownish-blond hair, blue eyes, and soft, round Russian faces. Neither could have been called glamorous. Nando was gangling, nearsighted, and somewhat shy; Susana, while youthful and sweet in appearance and with a fine figure, had an earnest, unflirtatious expression on her face.

While she drank her coffee, the flight was called. The Parrados, the two socialists, and everyone else in the restaurant went down to the departure lounge and then passed through customs and passport control and out onto the tarmac. There they saw the shining white plane which was to take them to Chile. They climbed up an aluminum ladder to the door at the front of the fuselage, filed into the confined cabin, and filled up the seats, which were placed in pairs on either side of the aisle.

At 8:05 A.M. the Fairchild, No. 571 of the Uruguayan Air Force, took off from Carrasco airport for Santiago de Chile, loaded with forty passengers, five crewmen, and their baggage. The pilot and commander of the plane was Colonel Julio César Ferradas. He had served in the Air Force for more than twenty years, had 5,117 hours of flying experience, and had flown over the treacherous *cordillera de los Andes* twenty-nine times. His copilot, Lieutenant Dante Hector Lagurara, was older than Ferradas but not as experienced. He had once had to parachute out of a T-33 jet and was now flying the Fairchild under the eyes of Ferradas to gain extra experience, as was the custom in the Uruguayan Air Force.

The plane he was flying—the Fairchild F-227—was a twin-engined turboprop manufactured in the United States and bought by the Uruguayan Air Force only two years earlier. Ferradas himself had flown it down from Maryland. Since then it had only logged 792 hours: by aeronautical standards it was as good as new. If there was any doubt in the pilots' minds, it did not concern the qualities of the plane but rather the notoriously treacherous currents of air in the Andes. Only twelve or thirteen weeks before, a four-engined cargo plane with a crew of six, half of whom were Uruguayans, had disappeared in the mountains.

The flight plan filed by Lagurara was to take the Fairchild direct from Montevideo to Santiago by way of Buenos Aires and Mendoza, a distance of around nine hundred miles. The Fairchild cruised at about 240 knots; it would therefore take them approximately four hours, the last half hour of which would be over the Andes. By leaving at eight, however, the pilots expected to reach the mountains before noon and avoid the dangerous postmeridional turbulence. All the same, they worried about the crossing, because the Andes, though less than a hundred miles wide, rose to an average height of 13,000 feet, with peaks as high as 20,000 feet; one mountain, Aconcagua, which lay between Mendoza and Santiago, rose to 22,834 feet, the highest mountain in the Western Hemisphere and only about 6,000 feet short of Mount Everest.

The highest the Fairchild could fly was 22,500 feet. It therefore had to fly through a pass in the Andes where the mountains rose to a lesser height. When visibility was good there was a choice of four—Juncal, the most direct route from Mendoza to Santiago, Nieves, Alvarado, or Planchon. If visibility was poor and the pilots had to fly on instruments, the Fairchild would have to go through Planchon, a hundred miles or so south of Mendoza, because Juncal had a minimum ceiling of 26,000 feet, and Nieves and Alvarado had no radio bea-

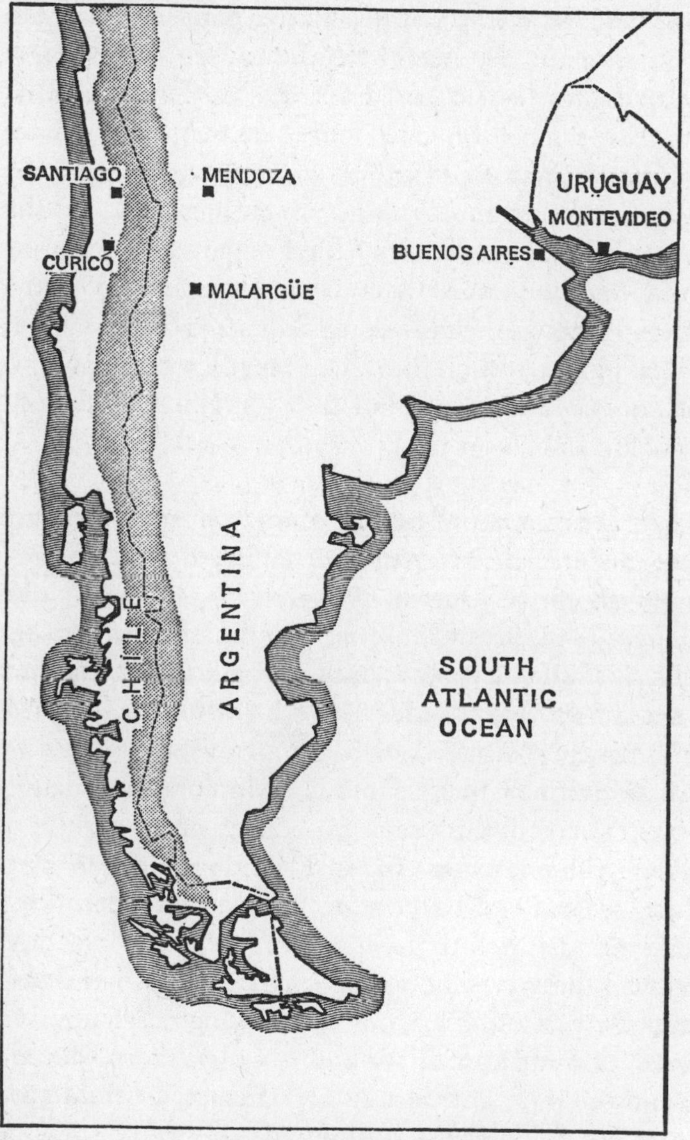

Southern South America

cons. The danger was not just that a plane might crash into a mountain. The weather in the Andes was subject to every kind of treachery. From the east, hot currents of air rose to meet the icy atmosphere at the snow line, which lay at between 14,000 and 16,000 feet. At the same time, the cyclonic winds which blew in from the Pacific roared up the valleys from the west and grappled in their turn with the hot and cold currents from the other side. If a plane was caught in such turbulence, it could be blown around like a leaf in a gutter. It was with these considerations in mind that Lagurara made contact with ground control at Mendoza.

There was no overt sign of anxiety in the passenger compartment. The boys talked, laughed, read comics, and played cards. Marcelo Pérez discussed rugby with members of the team; Susana Parrado sat next to her mother, who handed out candy to the boys around her. Behind them sat Nando Parrado with his greatest friend, Panchito Abal.

These two boys were famous as inseparable friends. They were both the sons of businessmen and both worked in their fathers' firms, Parrado marketing nuts and bolts, Abal in tobacco. On the surface it was an unequal friendship. Abal—handsome, charming, and rich—was one of the best rugby players in Uruguay and played for the Old Christians as a wing three-quarter; whereas Parrado was awkward, shy, and, though pleasant-looking, not particularly attractive. He played in the second line of the scrum.

The interests they had in common, besides rugby and business, were cars and girls, and it was this which had gained them the reputation of playboys. Cars are inordinately expensive in Uruguay, and both had one—Parrado a Renault 4 and Abal a Mini-Cooper. Both had motorcycles, too, which they took to Punta del Este and rode along the beaches with a girl on the pillion. Here again, their relationship seemed unequal, for while there was hardly a girl in Uruguay who did not want to be seen with Abal, a date with Parrado was not so popular. He lacked Abal's glamour and easy charm; moreover, he was neither more nor less than what he seemed to be. Abal, on the other hand, gave the impression that his gaiety and his easy, charming manner concealed a profound and mysterious melancholy which, together with an occasional expression of deep boredom, only added to his allure. Abal, in his turn, repaid the admiration of his female admirers with his time. His size, strength, and skill enabled him to skip some of the training that was essential for other members of the team, and what energies he saved from rugby were dedicated to these pretty girls, to cars and motorcycles, to elegant clothes for himself, and to his friendship with Parrado.

Parrado had one advantage over Abal for which the latter would willingly have exchanged all the others; he came from a happy, united family. Abal's parents were divorced. Both had been married before; both had had children by their previous marriages. His mother was much younger than his father, but it was with his father, who was now over seventy, that Abal had chosen to live. The divorce, however, had deeply injured him. His Byronic melancholy was not just an affectation.

* * *

The plane flew on over the endless pampas of Argentina. Those by the windows could see the geometric patches of green where crops had been planted amid the prairie and, every now and then, the squares of wood plantations, or small houses with trees planted around them. Then slowly the ground beneath them changed in appearance from a vast paving of green to the more arid ground at the foothills of the Sierras which rose to their right. The grass gave way to scrub and the cultivated land clung only to little spots which were artesian wells.

Suddenly they saw the Andes rising before them, a dramatic and apparently impassable wall with snow-clad peaks like the teeth of a giant saw. The sight of this, the cordillera, would have been enough to sober the most experienced traveler, let alone these young Uruguayans, most of whom had never seen mountains higher than the little hills which lie between Montevideo and Punta del Este. As they steeled themselves for the awesome sight of some of the highest mountains in the world, the steward, Ramírez, suddenly came out of the pilot's cabin and announced over a loudspeaker that weather conditions made it impossible to cross the cordillera. They were going to land in Mendoza and wait until the weather improved.

A groan of disappointment went up from the boys in the passenger compartment. As it was, they had only five days to spend in Chile and they did not want to waste one of them—or any of their precious U.S. dollars—in Argentina. However, since there was no way around the Andes, which run from one end of the South American continent to the other, there was nothing to be done, so they fastened their seat belts and sat tight as the Fairchild made a particularly rough landing at Mendoza airport.

When it had come to a halt by the airport building and Ferradas emerged from the pilots' cabin, a wing three-quarter on the team named Roberto Canessa somewhat impudently congratulated him on the landing.

"Don't congratulate me," said Ferradas. "Lagurara deserves the praise."

"And when are we going to Chile?" another boy asked.

The colonel shrugged his shoulders. "I don't know. We'll see what happens to the weather."

3

The boys followed the pilots and crew out of the plane and trooped across the tarmac to the customs control, the mountains of the precordillera brooding over them like an immense cliff face. Everything else was dwarfed—the buildings, the oil tanks, and the trees. The boys were undaunted. Not even the cordillera or the distasteful business of buying Argentinian pesos could suppress their high spirits. They left the airport and divided up into groups to take a bus or a taxi into town or to hitch a ride on some passing truck.

It was lunchtime and the boys were hungry. They had either had an early breakfast or no breakfast at all, and no food to speak of was carried on the Fairchild. A group of the younger ones made straight for a nearby restaurant, which they found was owned by an expatriate Uruguayan who would not let them pay for their meal.

Others went off in search of a cheap hotel and, having checked in, went out into the streets again to take a look around. Impatient as they were to get to Chile, they could not help but enjoy Mendoza. One of the oldest cities in Argentina, having been founded by the Spanish in 1561, it retained much of the grace and charm of the colonial period. Its streets were wide and lined with trees. The air, even so early in spring, was warm and dry that day, and scented with the new blossoms of the flowers planted in the public gardens. The streets were lined with pleasant shops, cafés, and restaurants, and outside the city were acres of vineyards producing some of the best wine in South America.

The Parrados, Abal, Señora Mariani, and the other two middle-aged couples booked themselves into one of the better hotels but went their different ways after lunch. Parrado and Abal found a motor race outside the city and in the evening joined Marcelo Pérez to see Barbra Streisand in *What's Up, Doc?* The younger boys met up with a group of Argentinian girls who were on a graduation holiday and took them dancing. Some did not get back to their hotels until four in the morning.

As a result, they slept late the next day. No word came from the crew that they should go to the airport, so once again they went wandering around the streets of Mendoza. One of the youngest boys, Carlitos Páez, who was something of a hypochondriac, stocked up on aspirins and Alka Seltzer. Others used the last of their Argentinian money to buy chocolate, nougat, and cartridges of butane gas to refill their lighters. Nando

Parrado bought a little pair of red shoes for his older sister's baby, and his mother bought small bottles of rum and liqueur for friends in Chile. She gave them to Nando to carry, and he stuffed them into an airline bag among his rugby clothes.

Two of the medical students, Roberto Canessa and Gustavo Zerbino, went to a café where tables and chairs were set out on the pavement of a tree-lined boulevard. There they ordered a breakfast of peach juice, croissants, and *café au lait*.

A short time later, as they sat drinking their coffee, they saw their captain Marcelo Pérez, and the two pilots walking toward them.

"Hey!" they shouted to Colonel Ferradas. "Can we leave now?"

"Not yet," said Ferradas.

"Are you cowards or what?" asked Canessa, who was nicknamed Muscles because of his stubborn character.

Ferradas, recognizing the high-pitched voice which had "congratulated" him on the landing the day before, looked momentarily annoyed. "Do you want your parents to read in the papers that forty-five Uruguayans are lost in the cordillera?" he asked.

"No," said Zerbino. "I want them to read, 'Forty-five Uruguayans cross the cordillera at all costs.' "

Ferradas and Lagurara laughed and walked on. They were placed in an awkward situation, not so much by the boys' taunts as by the dilemma they faced. The meteorological reports were that the weather was improving over the Andes. The pass at Juncal was still closed to the Fairchild, but there was an excellent chance that by the early afternoon Planchon would be clear. It would mean crossing the Andes at a time of day that was normally considered hazardous, but they were confident that they could fly above the turbulence. The only alternative was to return to Montevideo (because it was against regulations for a military aircraft of a foreign power to remain for more than twenty-four hours on Argentinian soil), which would not only disappoint the Old Christians but mean a serious loss of revenue for the impoverished Uruguayan Air Force. They therefore passed the word around through Marcelo Pérez that their passengers should report to the airport at one o'clock.

The passengers did as they had been told, but when they arrived they found no sign of the Uruguayan crew or the Argentinian officials who would have to check their baggage and passports. The boys fooled around while they waited, taking photographs, weighing themselves, frightening one another with the thought that it was Friday the thirteenth, and teasing Señora Parrado for taking a traveling rug to Chile in the spring. Then a cry went up. Ferradas and Lagurara had come into the airport building, both laden with large bottles of Mendoza wine. The boys began to tease them. "Drunks!" shouted one; "Smugglers!" shouted another; and the insolent Canessa said in a penetrating sneer, "Just look what kind of pilots we've got!"

Ferradas and Lagurara seemed a little disconcerted by the jeering crowd of boys. There was a latent defensiveness about them, partly because they were still undecided about what they should do and afraid that their caution would be mistaken for incompetence. Just at that moment, however, another plane landed at the airport. It was an old cargo plane which made a lot of noise and emitted spurts of smoke from its engines as it taxied over the tarmac, but when its pilot entered the airport building Ferradas approached him and asked for his advice.

The pilot had just come from Santiago and he reported that, though the air turbulence was strong, it should not prove a problem for the Fairchild, which had, after all, the most up-to-date navigational equipment. Indeed, this pilot even suggested that they take the direct route to Santiago over the Juncal Pass, which would reduce the journey to less than 150 miles.

Ferradas decided that they would go—not by the Juncal Pass but by the safer southerly route through the Pass of Planchon. A cheer went up from the boys when this was announced, though they still had to wait for the Argentinian officials to check their passports and let them through to the Fairchild.

In the meantime, they watched the dilapidated cargo plane take off again with the same noise and smoke as before. As it did so, two of the Old Christians turned to two Argentinian girls, who had gone out with them the night before and had now come to see them off, saying, "Now we know the kind of planes they have in Argentina."

"At least it got over the Andes," one of the girls replied tartly, "which is more than yours will."

4

The copilot, Lagurara, was again at the controls of the Fairchild as it took off from Mendoza airport at eighteen minutes past two, local time. He set his course for Chilecito and then Malargüe, a small town on the Argentine side of the Planchon Pass. The plane climbed to 18,000 feet and flew with a tail wind of between 20 and 60 knots.

The land beneath them was sparse and arid, marked by riverbeds and salt lakes which bore the tracery of bulldozers. To the right rose the cordillera, a curtain of barren rock reaching toward the sky. If the plain below was mostly infertile, these mountains were a desert. The brown, gray, and yellow rock was untouched by even the smallest trace of vegetation, for their height sheltered the mountains on this side from the rain which was blown in from the Pacific on the western side of the range. Here, on the Argentinian side, the soil which lay between the folds and cracks of the mountains was no more than volcanic dust. There were no trees, no scrub, no grass. Nothing broke the monotonous ascent of these brittle mountains except the snow. Above 13,000 feet it was perpetual, but at this time of the year it lay at much lower altitudes, softening the lines of the mountains and piling up in the valleys to a depth of more than a hundred feet.

The Fairchild was equipped not only with an ADF (Automatic Direction Finder) radio compass but also with the more modern VOR (VHF Omnidirectional Range). It was therefore a matter of routine to tune to the radio beacon at Malargüe, which was blocked at

15:08 hours. Still flying at 18,000 feet, the plane now turned to fly over the cordillera along air lane G 17. Lagurara estimated Planchon—the point in the middle of the mountains where he passed from Air Traffic Control in Mendoza to that in Santiago—at 15:21 hours. As he flew into the mountains, however, a blanket of cloud obscured his vision of the ground beneath. This was no cause for concern. Visibility above the clouds was good, and with the ground of the high cordillera covered with snow, there would, in any case, have been nothing by which they could have identified Planchon. Only one significant change had taken place: the moderate tail wind had now changed to a strong head wind. The ground speed of the plane had therefore been reduced from 210 to 180 knots.

At 15:21 Lagurara radioed Air Traffic Control in Santiago to say that he was over the Pass of Planchon and estimated to reach Curicó—a small town in Chile on the western side of the Andes—at 15:32. Then, only three minutes later, the Fairchild, once again made contact with Santiago and reported "checking Curicó and heading toward Maipú." The plane turned at right angles to its previous course and headed north. The control tower in Santiago, accepting Lagurara at his word, authorized him to bring the plane down to 10,000 feet as he came toward the airport of Pudahuel. At 15:30 Santiago checked the level of the Fairchild. It reported "level 150," which meant that Lagurara had already brought the plane down 3,000 feet. At this altitude it entered a cloud and began to jump and shake in the different currents of air. Lagurara switched on the sign in the passenger cabin which ordered passengers to fasten their safety belts and to stop smoking. He then instructed the steward, Ramírez, who had brought Ferradas a gourd filled with maté, the bitter tea of South America, to return to the galley and make sure that the unruly passengers did as they were told.

Inside the passenger compartment there was a holiday atmosphere. Several of the boys were walking up and down the aisle, peering out of the small windows to try and catch a glimpse of the mountains through a gap in the clouds. They were all in high spirits; they had their rugby ball with them, and some were throwing it up and down the passenger cabin over their heads. At the back of the plane a group was playing cards, and farther back still, by the galley, the steward and the navigator, Martínez, had been playing a game of *truco*, a kind of whist. As the steward made his way from the cockpit to resume the game, he told the boys still standing in the aisle to sit down.

"There's bad weather ahead," he said. "The plane's going to dance a little, but don't worry. We're in touch with Santiago. Well be landing soon."

As he approached the galley he told four of the boys at the back to move forward. Then he sat down with the navigator and took up his hand.

As the plane entered another cloud bank it began to shake and lurch in a manner which alarmed many of the passengers. There were one or two practical jokes to hide this nervousness. One of the boys took hold of the microphone at the back of the plane and said, "Ladies and gentlemen, please put on your parachutes. We are about to land in the cordillera."

His audience was not amused, because just at that moment the plane hit an air pocket and plummeted several hundred feet. Roberto Canessa, for example, feeling alarmed, turned to Señora Nicola, who sat with her husband across the aisle, and asked her if she was afraid.

"Yes," she said. "Yes, I am."

Behind them a group of boys started to chant "Conga, conga, conga," and Canessa, with a show of courage, took a rugby ball which he had in his hands and threw it to Dr. Nicola, who in his turn threw it back down the cabin.

Eugenia Parrado looked up from her book. There was nothing to be seen from the window but the white mist of cloud. She turned the other way and looked at Susana's face and took hold of her hand. Behind them Nando Parrado and Panchito Abal were engrossed in conversation. Parrado had not even fastened his seat belt, nor did he do so when the plane hit a second air pocket and sank like a stone for a further few hundred feet. A cry of "*¡Ole, ole, ole!*" went up from the boys in the cabin—those, that is, who could not see out of a window—for the second fall had brought the plane out of the clouds, and the view which opened up beneath them was not of the fertile central valley of Chile many thousands of feet below but of the rocky edge of a snow-covered mountain no more than ten feet from the tip of the wing.

"Is it normal to fly so close?" one boy asked another.

"I don't think so," his companion replied.

Several passengers started to pray. Others braced themselves against the seats in front of them, waiting for the impact of the crash. There was a roar of the engines and the plane vibrated as the Fairchild tried to climb again; it rose a little but then there came a deafening crash as the right wing hit the side of the mountain. Immediately it broke off, somersaulted over the fuselage, and cut off the tail. Out into the icy air fell the steward, the navigator, and their pack of cards, followed by three of the boys still strapped to their seats. A moment later the left wing broke away and a blade of the propeller ripped into the fuselage before falling to the ground.

Inside what remained of the fuselage there were screams of terror and cries for help. Without either wings or tail, the plane hurtled toward the jagged mountain, but instead of being smashed to pieces against a wall of rock it landed on its belly in a steep valley and slid like a toboggan on the sloping surface of deep snow.

The speed at which it hit the ground was around 200 knots, yet it did not disintegrate. Two more boys were sucked out the back of the plane; the rest remained in the fuselage as it careered down the mountain, but the force of deceleration caused the seats to break loose from their mountings and move forward, crushing the bodies of those caught between them and smashing the partition which separated the passenger cabin from the forward luggage area. While the freezing air of the Andes rushed into the decompressed cabin and those passengers who still had their wits about them waited for the impact of the fuselage against the rock, it was the metal and plastic of the seats which injured them. Realizing this, some of the boys tried to undo their safety belts and stand up in the aisle, but only Gustavo Zerbino succeeded. He stood with his feet planted firmly on the floor and his hands pressed against the ceiling, shouting, "Jesus, Jesus, little Jesus, help us, help us!"

Another of the boys, Carlitos Páez, was saying a Hail Mary, begun when the wing of the plane had first touched the mountain. As he mouthed the last words of this prayer, the plane came to a stop. There was a moment of stillness and silence. Then, slowly, from all over the tangled mess within the passenger cabin came the sounds of life—groans and prayers and cries for help.

As the plane had hurtled down the valley, Canessa had braced himself for the impact, thinking that in a moment he would die. He did not pray but calculated in his mind the speed of the plane and the force with which it would hit the rock. Then suddenly he was conscious that the plane was no longer moving.

He shouted, "It's stopped!" and then turned to the boy who sat beside him and asked him if he were all right. The boy was in a state of shock. He nodded and Canessa left him to help his friend Daniel Maspons extricate himself from his seat. Then the two of them started to help others. At first they thought that they were the only two who were not injured, for all around them they could hear cries for help, but others began to emerge from the wreckage. First came Gustavo Zerbino, then the team captain, Marcelo Pérez. Pérez had a bruised face and a pain in his side, but as captain of the team he immediately took it upon himself to organize the rescue of those trapped in the wreckage, while two medical students, Canessa and Zerbino, did what they could for the injured.

Immediately after the plane had stopped, some of the younger boys, smelling the gasoline fumes and fearing that the plane might explode or catch fire, had jumped out the gaping hole at the back. They found themselves up to their thighs in snow. Bobby Francois, the first to leave the plane, climbed onto a suitcase and lit a cigarette. "We've had it," he said to Carlitos Páez, who had followed him out into the snow.

The scene was one of utmost desolation. All around them was snow and beyond, on three sides, the sheer gray walls of the mountains. The plane had come to a halt on a slight tilt, facing down the valley where the mountains were much farther away and now partly obscured by gray clouds. It was bitterly cold, and many of the boys were in their shirtsleeves. Some wore sports coats and others blazers. None was dressed for subzero temperatures, and few suitcases could be seen which might provide extra clothes.

As they looked back up the mountain for their lost luggage, this group of younger boys saw a figure staggering down the mountainside. As it drew nearer, they recognized one of their friends, Carlos Valeta, and shouted to him, calling him to come in their direction. Valeta seemed unable to see or hear them. At each step he sank up to his thighs in the snow, and only the steepness of the hill enabled him to make any progress at all. The boys could see that his course would not lead him to the plane, so they shouted yet more frantically to attract his attention. Two, Páez and Storm, even tried to go out to meet him, but it was impossible to walk in the snow, particularly uphill. They were trapped and could only watch helplessly as Valeta stumbled down into the valley. For a moment it seemed as if he might have heard them and was changing direction toward the plane, but then he slipped. His wide stride became a tumble, and his body slithered helplessly down the side of the mountain until he finally disappeared in the snow.

Inside the plane, the handful of boys who were able tried to prize away the seats which trapped so many of the wounded. In the thin air of the mountains it took double the energy and effort, and those who had suffered only superficial injuries were still in a state of shock.

Even when the wounded were pulled clear, there was little that anyone could do. The training of the two "doctors," Canessa and Zerbino (a third medical student, Diego Storm, was in a state of shock), was pitifully inadequate. Of Zerbino's first year at medical school, six months had been dedicated to compulsory classes in psychology and sociology. Canessa had done two years, but even so this was only a quarter of the total course. All the same, both were aware of the special responsibility which their training conferred upon them.

Canessa knelt down to examine the crushed body of a woman which at first he could not recognize. It was Eugenia Parrado, and she was dead. Beside her lay Susana Parrado, who was semiconscious and alive but seriously injured. Blood poured out of a gash in her forehead and blinded one eye. Canessa wiped away the blood, so that she could see, and then laid her down on the small part of the floor that was not cluttered with seats.

Near her was Abal. He too was severely injured, with an open wound in his scalp. He was semiconscious and, as Canessa knelt to treat him as best he could, Abal took hold of his hand, saying, "Please don't leave me, old man, please don't leave me." There were so many others crying for help that Canessa could not stay with him. He called to Zerbino to tend to Abal and moved on to Parrado, who had been thrown out of his seat and lay senseless at the front of the plane. His face was covered with bruises and blood and Canessa thought that he was dead. He knelt and felt for a pulse; a faint beating of the heart registered on his fingertips. Parrado was still alive, but it seemed impossible that he could live for long, and since nothing could be done to help him, he was given up for dead.

Besides Eugenia Parrado, only two other passengers in the fuselage had died instantly. These were the Nicolas; both had been flung forward into the luggage compartment side by side and had died at once.

For the time being their bodies were left where they were, and the two medical students returned to do what they could for the living. They made bandages of the antimacassars from the backs of the seats, but for many of the injuries these were quite inadequate. One boy, Rafael Echavarren, had had the calf of his right leg torn off and twisted around to cover the shin. The bone was entirely exposed. Zerbino took hold of the bleeding muscle, pulled it around to its proper place, and then bound up his leg with a white shirt.

Another boy—Enrique Platero—came up to Zerbino

with a steel tube sticking into his stomach. Zerbino was appalled, but he remembered from his lessons in medical psychology that a good doctor always instills confidence into his patient, so he looked Platero straight in the eye and, with as much conviction as he could put into his tone of voice, said, "Well, Enrique, you look all right."

"Do you think so?" said Platero, pointing to the piece of steel in his stomach. "And what about this?"

"Don't worry about that," said Zerbino. "You're perfectly strong, so come and give me a hand with these seats."

Platero seemed to accept this. He turned toward the seats and, as he did so, Zerbino grabbed hold of the tube, put his knee against Platero's body, and pulled. The piece of steel came out, and with it came almost six inches of what Zerbino took to be Platero's intestine.

Platero, his attention once more upon his stomach, contemplated his projecting innards with some dismay, but before he had time to complain, Zerbino said to him, "Now look here, Enrique, you may think you're in a bad way, but there are plenty of others much worse off than you are, so don't be a coward, and just come and help. Tie that up with a shirt, and I'll see to it later."

Without complaint Platero did as Zerbino had told him.

Canessa, meanwhile, had returned to Fernando Vázquez, the boy who had been sitting beside him. The leg Canessa had thought was merely broken had, in fact, been cut in two by the plane's propeller as it ripped into the fuselage. Blood had poured out of the severed artery. Now Vázquez was dead.

Many of the other boys had been injured in the legs as the seats buckled up and pressed together. One had his leg broken in three different places, was severely wounded in the chest, and was now unconscious. It was those who were conscious who suffered—Panchito Abal, Susana Parrado, and, worst of all, the middle-aged lady whom none of them knew, Señora Mariani. She was trapped by her two broken legs under a pile of seats, and the boys were unable to extricate her. She screamed and begged for help, but it was beyond their strength to lift the seats which held her down.

The face of Liliana Methol, the fifth woman in the plane, was badly bruised and covered with blood, but all her injuries were superficial. Her husband, Javier, a cousin of Panchito Abal, was unhurt, but altitude sickness had come over him with the virulence of influenza. Though he made feeble attempts to help the wounded, he felt such dizziness and nausea that he was barely able to move. Others, though not afflicted with the same symptoms, suffered from the shock of the accident. One boy, Pedro Algorta, had total amnesia. He was physically well enough to work hard at moving the seats, but he had no idea of where he was or what he was doing. Another boy had also been hit on the head and made repeated efforts to leave the plane and walk down the mountain.

The plane had crashed at about half past three in the afternoon. Because of the clouds the light was already somber, and around four o'clock it began to snow. The few flakes which fell at first grew into a flurry and then fell thickly, obliterating the view of the mountains. In spite of the snow, Marcelo Pérez directed that the wounded should be carried out so that those who were fit could clear the tangled seats from the floor of the Fairchild. This was intended as a temporary measure; they all felt sure that by now the plane would be reported missing and help would be on the way.

They realized that the rescue might be made easier if they could transmit signals from the radio. The entrance to the pilots' cabin was blocked by the wall of seats which had piled up at the top of the passenger compartment, but sounds of life could be heard on the other side, so one of the calmer boys, Moncho Sabella, decided to try to reach the pilots from the outside.

It was almost impossible to walk on the snow, but he discovered that he could use seat cushions as stepping-stones to the front of the plane. The nose had been crushed by the descent, but it was not difficult to climb up and look into the cockpit through the door to the front luggage compartment.

There he discovered that Ferradas and Lagurara were trapped in their seats, with the instruments of the airplane embedded in their chests. Ferradas was dead, but Lagurara was alive and conscious and, seeing Sabella beside him, begged him to help. There was little Sabella could do. He could not move Lagurara's body, but in answer to his plea for water he crammed some snow into a handkerchief and held it to his mouth. After that he tried to make the radio work but it was completely dead; when he returned to the others, however, to keep up their morale he told them that he had spoken with Santiago.

Later, Canessa and Zerbino retraced Sabella's steps

to the pilots' cabin. They tried to push the instrument panel off Lagurara, but it was impossible to move it even half an inch. His seat was also fixed immovably in position. All they were able to do was remove the cushion at the back and thereby relieve some of the pressure on his chest.

As they struggled in this futile attempt to free him, Lagurara said over and over again, "We passed Curicó, we passed Curicó." Then, seeing that nothing could be done, he asked the two boys to fetch the revolver which he kept in his bag. The bag was nowhere to be seen, nor would Canessa and Zerbino have given him the gun if they had found it, because, as Catholics, they could not condone suicide. They asked him instead if they could use the radio to bring help and set the dial as Lagurara instructed, but the transmitter was dead.

Lagurara continued to beg for his revolver and then asked for water. Canessa climbed out of the cockpit and brought in some snow which he fed into Lagurara's mouth, but the man's thirst was pathological and insatiable. He was bleeding through the nose, and Canessa knew he would not live for long.

The two "doctors" made their way back over the seat cushions to the rear of the plane and returned to the dark, narrow tunnel of moaning, screaming humanity. Those who had been extracted from the wreckage lay out on the snow, as the few who were fit and strong worked desperately to drag out those seats they could prize loose and clear some space on the floor of the plane. But the daylight was fading. By six it was almost dark and the temperature had sunk far below freezing. It was clear that rescue would not come that day, and so the wounded were brought back into the plane and the thirty-two survivors prepared for the night.

5

There was little space in which anyone could stand, let alone lie down. The break at the back of the fuselage was jagged, leaving seven windows on the left-hand side of the plane but only four on the right. The distance from the pilots' cabin to the gaping hole at the rear measured only twenty feet, and most of this space was taken up by the knotted tangle of seats. The only floor space they had been able to clear before dark was by the entrance, and it was here that they laid the most seriously wounded, including Susana and Nando Parrado and Panchito Abal. In this position they were able to lie almost horizontally, but they had little protection against the snow and the bitter wind that blew in from the darkness. Marcelo Pérez, with the help of a hefty wing forward named Roy Harley, had done his best to build a barrier against the cold with anything that came to hand—especially the seats and suitcases—but the wind was strong and their wall kept falling down.

Pérez, Harley, and a group of uninjured boys remained in a huddle near the wounded by the entrance, drinking the wine which the pilots had bought in Mendoza and doing what they could to keep up their makeshift barrier. The rest of the survivors slept where they could among the seats and bodies. As many as possible, including Liliana Methol, moved into the confined space of the luggage compartment, which lay between the passenger cabin and the cockpit. It was cramped and uncomfortable but by far the warmest place in the plane. There, too, they passed around the large bottles of Mendoza wine. Some of the boys were still dressed in short-sleeved shirts, and they gulped down quart after quart to bring some warmth to their bodies. They also pummeled and massaged one another. This seemed the only way to keep warm, until Canessa had the first of his ingenious ideas. He found, by examining the cushions and seats which lay all around them, that the upholstery, which was turquoise and made of brushed nylon, was only held to the seats by a type of zipper. It was quite simple to remove the coverings and, once removed, these made small blankets. They were pitifully inadequate protection against subzero temperatures, but they were certainly better than nothing.

Worse than the cold that night was the atmosphere of panic and hysteria in the cramped cabin of the Fairchild. Everyone thought that his injury was the worst and complained out loud to the others. One boy who had broken his leg screamed at anyone who came near him. He said that they were hurting his leg and cursed them for it, but when he himself wanted to go to the entrance to scoop up some snow to quench his thirst, he would clamber over others with complete disregard for their injuries. Marcelo Pérez did what he could to calm him. He also tried to cope with Roy Harley, who became hysterical every time part of their wall fell down.

All the time there came from the dark the moans, screams, and delirious raving of the wounded. In the luggage compartment they could still hear cries and groans from Lagurara; "We passed Curicó," he would say, "we passed Curicó." He would moan for water and beg for his gun.

Inside the cabin itself the worst cries came from Señora Mariani, still trapped by her broken legs beneath the seats. At one time an attempt was made to free her, but it was impossible. As they worked, her cries became shriller still, and she swore that if they moved her she would die. They gave up the attempt. Two boys, Rafael Echavarren and Moncho Sabella, took hold of her hand in an attempt to comfort her, and to some extent they succeeded, but later the shrieks continued.

"For God's sake, be quiet!" came cries from the back of the plane. "You're no worse off than anyone else."

At this the cries redoubled.

"Shut up," shouted Carlitos Páez, "or I'll come and smash your face in!"

Her screaming grew louder and more insistent, then it abated, but then it started again as a boy who was still in a state of shock stepped on her as he tried to get to the door.

"Keep him away!" she shouted. "Keep him away, he's trying to kill me. He's trying to kill me!"

The "assassin," Eduardo Strauch, was pulled back onto the floor by his cousin, but a short time later he stood up and tried to climb over the seats and bodies to find a warmer and more comfortable place to sleep. This time he stood on the only surviving member of the crew (besides Lagurara)—Carlos Roque, the mechanic. He too took Eduardo Strauch for a murderer and, with the punctiliousness of a trained serviceman, asked him to identify himself.

"Show me your papers!" he shouted. "Identify yourself. Identify yourself."

When Eduardo did not present his passport but continued to clamber over Roque toward the door, the mechanic became hysterical.

"Help me!" he screamed. "He's mad. He's trying to kill me." Once again Eduardo was pulled back into place by his cousin.

In another part of the plane a second figure, Pancho Delgado, stood up and made for the door. "I'm just going to the store to get some Coca-Cola," he announced to his friends.

"Then get me some mineral water while you're there," replied Carlitos Páez.

In spite of the intense discomfort, some of the boys managed to drift off to sleep, but it was a long night. The cries of pain continued as one boy stumbled over broken limbs to scoop up snow at the entrance or another awoke, not knowing where he was, and tried to leave the plane. There were cries, too, from those who were irritated by the bleatings of self-pity, and there were acrimonious exchanges between some Old Christians and the boys from the Jesuit College of the Sacred Heart.

Those who were awake huddled closer together as the wind blew through their improvised ramparts and the other holes that had been made in the wall of the fuselage. The boys by the entrance suffered most—their limbs numb with cold, their faces tickled by snowflakes which blew in over their bodies. The uninjured could at least buffet one another and rub their feet and fingers to keep up the circulation of their blood. It was the plight of the two Parrados and Panchito Abal that was most terrible. They were unable to warm themselves and, though the injuries they had sustained were so terrible, only Nando was unconscious, oblivious of his agony. Abal begged for help which no one could give him—"Oh, help me, please help me. It's so cold, it's so cold . . ."—and Susana cried continuously for her dead mother—"Mama, Mama, Mama, let's go away from here. Let's go home." Then her mind wandered and she sang a nursery rhyme.

In the course of the night the third medical student, Diego Storm, decided that, though Parrado was unconscious, his injuries seemed more superficial, so he pulled Parrado's body over among the group of his friends, and together they combined to keep it warm. It seemed senseless to do this for the other two.

The night was unending. At one point Zerbino thought he saw the dim light of dawn through their makeshift wall. He looked at his watch; it was only nine o'clock at night. Later still, those in the middle of the plane heard a foreign voice at the entrance. For a moment they thought it was a rescue party, but then they realized that it was Susana, praying in English.

6

The sun rose on the morning of Saturday, October 14, to find the hulk of the Fairchild half buried in snow. It lay at about 11,500 feet, between the Tinguiririca volcano in Chile and the Cerro Sosneado in Argentina. Though the plane had crashed more or less in the middle of the Andes, its exact position was on the Argentine side of the border.

The plane lay on a slant. Its bent nose pointed down the valley, which fell away steeply toward the east. In every other direction, beyond the carpet of deep snow, arose walls of immense mountains. Their slopes were not sheer; rather, they spread themselves, huge and inhospitable. Occasionally, the gray and pink of the rough volcanic rock appeared through the snow, but at that altitude nothing grew in the shale—no bush, no scrub, no blade of grass. The plane had crashed not just in mountains but in a desert, too.

The first out of the plane were Marcelo Pérez and Roy Harley, who pushed down the barricade they had been at such pains to erect the night before. There were clouds in the sky but the snow had stopped. The frost had frozen the surface of the snow, and they were able to take some steps away from the plane and study the utter bleakness of the situation.

Inside the plane Canessa and Zerbino began, once again, to examine the wounded and discovered that three more had died in the night, including Panchito Abal, who lay motionless over the body of Susana Parrado. His feet were blackened by frostbite, and it was quite evident from the stiffness of his limbs that he was dead. For a time they thought that Susana must also be dead for she was quiet, but when they removed Abal's body they saw that she was still alive and still conscious. Her feet, too, had gone purple from the cold, and she complained to the mother who was no longer there. "Mama, Mama," she cried, "my feet are hurting. They are hurting so much. Oh, please, Mama, can't we go home?"

There was little that Canessa could do for Susana. He massaged her frostbitten feet to try and bring back the circulation, and again he wiped away the caked blood from her eyes. She was sufficiently conscious to be glad that she had not gone blind, and she thanked Canessa for his care. Canessa was only too well aware that the superficial cuts on her face were, in all probability, the least of her injuries. He felt sure that her internal organs were badly damaged, but he had neither the knowledge nor the facilities to do anything about it. Indeed, there was little he could do for any of them. There were no drugs on the plane beyond those which Carlitos Páez had bought in Mendoza and some Librium and Valium found in a handbag. There was nothing in the wreckage which seemed suitable to be used as splints for broken limbs, so Canessa could only tell those with broken arms or legs to lay them on the snow to help bring down the swelling; later he advised them to massage their sprained ligaments. He was afraid to tie the bandages he had made from the antimacassars too tightly because he knew that in the extreme cold this might impede the circulation of the blood.

When he came to Señora Mariani, he thought that she too was dead. He crouched down beside her and made another attempt to move the seats which still pinned her to the floor, whereupon she started to scream once again. "Don't touch me, don't touch me! You'll kill me!" so he decided to leave her alone. When he returned later in the morning to see how she was, she had taken on a lost look and was silent. And then, just as he looked into her eyes, they rolled around into her skull and she ceased to breathe.

Canessa, though he had studied medicine for a year longer than Zerbino, could not bring himself to say that someone was definitely dead. He left it to Zerbino to kneel and place his ear to Señora Mariani's chest to listen for the faintest sound of her heart. There was none, and so the other boys pushed back the seats, tied a nylon belt from the luggage hold around the shoulders of the corpse, and dragged it out into the snow. They told Carlitos Páez that she was dead, and Páez was filled with remorse for his harsh words of the night before and hid his face in his hands.

It was also Gustavo Zerbino who examined the hole in Enrique Platero's stomach from which he had pulled the steel tube the day before. He unwound the shirt and there, just as he had feared, was a protruding piece of gristle which so far as he knew could be part of the intestine or the lining of the stomach. It was bleeding, and to stem the flow of blood Zerbino tied it up with some thread, disinfected it with eau de cologne, and then told Platero to push it back into his stomach and bind up the wound once again. This Platero did without complaint.

The two doctors were not without their nurse. Liliana Methol, though her face was still purple from the bruises that she had received in the crash, did what she could to help and encourage them. She was a small, dark woman whose whole life until then had been devoted to her husband, Javier, and four children. Before they had married, Javier had suffered an accident. He had been knocked off a motor bicycle and then run over by a car; as a result, he had been unconscious for several weeks and in the hospital for several months after that. His memory never completely recovered, and he had lost the use of his right eye.

This was not his only misfortune. When he was twenty-one his family had sent him to Cuba and then to the United States, to study the production and marketing of tobacco. In the town of Wilson, North Carolina, he was told that he had tuberculosis. His illness was too far advanced for him to return to Uruguay, and he had to spend the next five months in a North Carolina sanatorium.

Even after his return to Montevideo he was bedridden for four more months, but there he would be visited by his girlfriend, Liliana. He had known her since he was twenty, and on June 16, 1960, they were married. For their honeymoon they went to Brazil, but only once since then had they been abroad—to the lakes of southern Argentina. It was as a belated celebration of their twelfth wedding anniversary that Javier had brought Liliana on the trip to Chile.

After the accident Liliana had been the first to notice that Javier—almost alone among the survivors—suffered chronically from the altitude. He remained nauseous and weak. His movements were heavy and his mind became slow. Liliana had to show him where to go and what to do and strengthen his spirit with her resolve.

She was also the natural source of comfort for the younger boys. Many of them were not yet twenty. Many of them, too, had been cared for throughout their lives by admiring mothers and sisters, and in their terror and despair they turned to Liliana, who—aside from Susana—was the only woman among them. She answered their need. She was patient and kind, speaking soft words to make their spirits strong. When, on the first night, Marcelo and his friends insisted that she sleep in the warmer part of the plane, she accepted their chivalrous gesture, but the next day she insisted that she be treated in just the same way as the others. Though some of the younger boys such as Zerbino would still have liked to accord her some deference and privacy, they had to recognize that in the confines of the plane the segregation of the sexes was not practicable, and from then on she was treated as one of the team.

The attention of the doctors and their nurse was drawn to one of the youngest members of the first fifteen, Antonio Vizintín, called Tintin, who seemed to be suffering from concussion and had been put to rest on the webbing of the luggage area. It was only now, the day following the accident, that blood was seen dripping from the sleeve of his coat. When they asked Vizintín what was wrong with his arm, he insisted that it was perfectly all right because he felt no pain. Liliana took a closer look, however, and found that the sleeve of his jacket was sodden with blood. The two doctors were summoned and, finding it impossible to remove Vizintín's jacket, they cut away the sleeve with a penknife. As they pulled it away—heavy with the blood that had soaked into it—more blood gushed from a severed vein. They at once made a tourniquet to stop the flow from his body and then bandaged the wound as best they could. Vizintín still felt no pain but he was very weak. Skeptical about his chances of survival, Canessa and Zerbino laid him down again on his bed in the luggage compartment.

Their final tour of duty was to the pilots' cabin. Since

early that morning no sound had come from Lagurara, and when they forced their way in from the luggage compartment they found, as they had suspected, that he was dead.

With the death of Lagurara they had lost the one man who might have told them what they should do to facilitate their rescue, for Roque, the only surviving crew member, was of little use. Since the crash he had wept continuously and lost all control of his bodily functions, being conscious of soiling his trousers only because of the complaints of those around him and the actions of the boys who helped to change them.

He was, all the same, in the Uruguayan Air Force, and Marcelo Pérez asked him if there were any emergency supplies or signal flares in the plane. Roque said no. Marcelo then asked him if the radio could be made to work, and Roque replied that it would need the power of the plane's batteries, which had been stored in the missing tail.

There seemed nothing to be done, but Marcelo was so confident they would soon be rescued that he was not unduly concerned. It was agreed, all the same, that what food they had should be rationed, and Marcelo made an inventory of everything edible that had been salvaged from the cabin or from those pieces of luggage which had not been lost with the tail. There were the bottles of wine which the pilots had bought in Mendoza, but five of them had been drunk in the course of the night and only three remained. There was also a bottle of whiskey, a bottle of cherry brandy, a bottle of crème de menthe, and a further hip flask of whiskey, half of which had been drunk.

For solid food they had eight bars of chocolate, five bars of nougat, some caramels which had been scattered over the floor of the cabin, some dates and dried plums, also scattered, a packet of salted crackers, two cans of mussels, one can of salted almonds, and a small jar each of peach, apple, and blackberry jam. This was not a lot of food for twenty-eight people, and since they did not know how many days they would have to wait before being rescued, it was decided to make it last as long as possible. For lunch that day Marcelo gave each of them a square of chocolate and the cap from a deodorant can filled with wine.

That afternoon they heard an airplane flying overhead but saw nothing because of the clouds. Again night came upon them more quickly than expected, but this time they were better prepared. They had cleared more space in the plane, they had built a better wall against the wind and snow, and there were fewer of them.

7

On the morning of Sunday, October 15, those who came out of the plane saw that for the first time since they had crashed the sky was clear. It was a deep blue, quite unlike any sky they had seen before, and in spite of their circumstances the survivors were impressed by the grandeur of their silent valley. The surface of the fresh snow had frozen, the crystals reflecting the bright, unfiltered sun. All around them were mountains, dazzling now in the bright light of early morning. Distances were deceptive. In the thin air, the peaks seemed close at hand.

The clear skies gave them reason to believe that they would be rescued that day, or at least be spotted from the air. In the meantime there were certain problems facing them, and they set about their solution in a more methodical way. Their most pressing need was for water. The snow was difficult to melt in sufficient quantity to quench their thirst, and to eat it merely froze one's mouth. They found that it was better to compress it into a ball of ice, and then suck it, or cram it into a bottle and shake the bottle until the snow melted. This latter process, however, not only took time but energy and provided barely enough for the needs of a single person. There were also the wounded who were not well enough to help themselves: Nando and Susana Parrado and Vizintín, who craved water to replenish the blood he had lost from the wound in his arm.

It was Adolfo Strauch who invented a water-making device. Adolfo—or Fito, as he was called—was an Old Christian, but he played for a rival rugby team in Montevideo. He had been persuaded to come to Chile at the last moment by Eduardo Strauch, his double cousin (their fathers being brothers, their mothers sisters). The Strauch family had come to Uruguay from Germany in the nineteenth century and had built up large business interests in banking and the soap industry. Fito and Eduardo came from a junior branch of the family; their fathers were trained as jewelers and until recently had been partners in Montevideo. On their mothers' side, the two boys came from the well-known Uruguayan family of Urioste.

Both boys were blond and good-looking, with recognizably German features. In fact, Eduardo was called "the German" by the others. The two were extremely close, more brothers than cousins, but while Eduardo had already developed serious ambitions as an architect, and had traveled to Europe, Fito was shy and indecisive about his life. He was a student of agronomy because he had no vocation for anything else and his family owned a ranch. Until this trip, he had never set foot out of Uruguay.

In the crash both Fito and Eduardo had been knocked unconscious. When they regained consciousness they were in such a state of shock that they did not realize where they were. Fito had tried to leave the plane immediately, and during the first night it had been Eduardo who had stepped on Roque and Señora Mariani. They had been restrained by another cousin who had survived, Daniel Fernández, the son of their fathers' sister.

By Sunday, Fito had recovered sufficiently to apply his mind to the difficulty they were facing in converting snow into water. The sun shone brightly, and as the morning proceeded its rays became increasingly hot, melting the brittle crust on the surface of the snow that had formed the night before. It occurred to Fito that they might somehow harness this solar heat to make water. He looked around for something to hold the snow, and his eye fell upon a rectangle of aluminum foil, measuring about one by two feet, which came from inside the back of a smashed seat thrown from the plane. He extracted it from the upholstery, bent up the sides so that it formed a shallow bowl, and twisted one corner to make a spout. He then covered it thinly with snow and tilted the whole apparatus to face the sun. In a short time drops of water appeared in the spout, and then a steady trickle poured into the bottle that Fito held ready beneath it.

Since every seat contained such a rectangle of aluminum, there were soon several water makers at work. Melting snow required a minimum of physical energy, so this became the regular task of those who were not well enough to do anything more strenuous, for Marcelo had decided to organize the survivors into different groups. He took upon himself the position of general coordinator and distributor of the food. The first group was the medical team, composed of Canessa, Zerbino, and to a lesser extent Liliana Methol. (Its composition was somewhat vague because Canessa refused to be limited by any narrow definition of his function.) The second group were those who were put in charge of the cabin. It was made up of some of the younger boys, such as Roy Harley, Carlitos Páez, Diego Storm, and the central figure in that group of friends, Gustavo Nicolich, whom they called Coco. It would be their duty to keep the cabin tidy, to prepare it at night by laying cushions on the floor, and in the morning to dry in the sun the covers of the seats which they had used as blankets.

The third team was the water makers. Their only difficulty was in finding uncontaminated snow, for around the plane it was pink from the blood of the dead and wounded and polluted by oil from the plane and by urine. There was no shortage of pure snow a few yards away, but either it was so soft that it was impossible to walk those few yards or, as in the early morning when the surface was frozen, it was hard enough to bear their weight but too hard to scoop up onto the aluminum pans. It was therefore decided that two areas only should be used as lavatories—one just by the entrance and the other beneath the front wheel of the plane underneath the pilots' cabin.

At midday Marcelo was to give out the ration of food. For each there was to be a deodorant cap filled with wine and a taste of jam. A square of chocolate would be kept for the evening meal. There were some complaints that there should have been a little bit more for their Sunday lunch, but the majority agreed that it was better to be careful.

There was now one more to share the rations. Parrado, once left for dead, had regained consciousness that day, and when the blood was washed from his face it was found that most of it had come from the blow on his head. His skull was intact, but he was weak and somewhat confused. His first thought had been for his mother and sister.

"Your mother died at once in the crash," Canessa told him. "Her body is out in the snow. But don't think of that. You must help Susana. Rub her feet and help her to eat and drink."

Susana's condition had deteriorated. Her face was still covered with cuts and bruises, and worse still her feet were now black from that first night. She was usually conscious yet seemed unaware of where she was. She still cried for her mother.

Nando massaged her frostbitten feet, but it was useless. No warmth returned, and when he rubbed them harder the skin came away in his hand. From then on he devoted himself to her care. When Susana mumbled that she was thirsty, Nando held to her lips the mixture of snow and crème de menthe and the little pieces of chocolate which Marcelo had set aside for her. And when she murmured, "Mama, Mama, I want to go to the bathroom," he rose and went to consult Canessa and Zerbino.

They came to her and Nando told her they were doctors.

"Oh, doctor," said Susana, "I would like a bedpan."

"But you have one," Zerbino answered. "Just go to the bathroom. It will be all right."

Shortly after noon, the boys saw a plane flying directly overhead. It was a jet and flying high above the mountains, but all those who were out in the snow jumped up, waved, shouted, and flashed pieces of metal up into the sky. Many cried with joy.

At midafternoon a turboprop flew over them from east to west, this time at a much lower altitude, and soon after that another flew over them from north to south. Again the survivors waved and shouted, but the plane continued on its course and disappeared over the mountains.

There now arose a division among the boys as to whether they had been seen or not, and to settle the dispute they turned to Roque, who was emphatic that a plane flying so low would have seen them.

"Then why didn't it circle," asked Fito Strauch, "or dip its wings to let us know that we'd been seen?"

"Impossible," said Roque. "The mountains are too high for that sort of maneuver."

The skeptics among them did not trust Roque's judgment; his behavior was still irrational and infantile at times and his optimism only made them more doubtful. Some of them began to think that perhaps the hulk of the Fairchild, with its roof half buried in the snow, might be more difficult to spot from the air than they had imagined. As a result they started to paint a scarlet SOS on the roof of the plane with lipstick and nail polish from the women's handbags, but as soon as they had completed the first S they realized that it would be much too small to make any difference.

Then, at half past four, they all heard the engines of an airplane much nearer to them than ever before, and there appeared from behind the mountains a small biplane following a course which would pass directly over them. They waved frantically and tried to reflect the sun up into the eyes of the pilot with their small pieces of metal, and to their intense joy the biplane, as it flew over them, dipped its wings as if signaling that they had been seen.

Nothing now could stop the boys believing what they wanted so much to believe, and while some simply sat in the snow waiting for the arrival of helicopters, Canessa opened a bottle of the Mendoza wine and with the wounded who were in his charge gulped it down to celebrate their salvation.

Shortly afterward it began to grow dark. The sun went behind the mountains and the bitter cold returned. No sound broke the silence. It was clear that they would not be rescued that night. Marcelo gave out the ration of chocolate to the survivors, who then shuffled back into the plane, some jockeying for position to avoid sleeping near the entrance. Marcelo pleaded with the strong to sleep with him in the cold, but several refused to give up the territory they had claimed in the warmer luggage area, saying that if they slept by the entrance every night they would certainly freeze to death.

For a long time that Sunday night no one slept. They talked of rescue—some arguing that helicopters would come the next day, others contending that they were too high for helicopters, that rescue might take longer, perhaps even a week. It was a sobering thought, and Canessa was rebuked severely by Marcelo for the thoughtless gluttony with which he and his patients had consumed a whole bottle of wine. And there was one among them who would have been more severely rebuked still if his identity had been known, for Marcelo had discovered that two pieces of chocolate and one bar of nougat had been stolen from the vanity case in which he kept their supplies.

"For God's sake," said Marcelo, pleading rhetorically with the unknown thief. "Don't you realize that you're playing with our lives?"

"The son of a bitch is trying to kill us," said Gustavo Nicolich.

It was cold and dark. They lapsed into silence, and each boy was left alone with his thoughts. Parrado slept with Susana clasped in his arms, enveloping her with his tall body to give her all the warmth he could, aware of her fitful breathing, the irregular movement in and out of her lungs, broken by cries for her dead mother. When he was able to look into Susana's large eyes he saw all the sorrow, pain, and confusion that she could not express in words. Others, too, slept only fitfully, clutching the makeshift blankets around them. Confined in a space measuring twenty feet by eight, they could fit in only by lying sandwiched together in couples, end to end, the feet of one resting on the shoulders of another. The plane itself was tilted on its axis. Those who lay full length on the floor were at an angle of about 30 degrees. Those opposite had only their legs on the floor, resting their backs against the wall of the cabin and their buttocks on the overhead rack, which they had torn down and used as a slat to break the angle between the floor and the wall.

Though there was some comfort provided by the cushions, the space was so cramped that for one to move his position meant that all had to do the same. Any movement whatsoever caused agony to those with broken legs, and the poor wretch who wished to scratch himself or urinate precipated a flood of abuse from those around him. Often their movements were involuntary. A boy would move his leg in his sleep and in so doing kick the face of the one opposite. Every now and then one of their number would sleepwalk. "I'm off to get some Coca-Cola!" he would shout, and then start to climb over the bodies which lay between him and the door.

No one reacted more irritably to the disturbances than Roberto Canessa, stung with remorse at the wine episode and by nature high-strung and volatile. Psychological testing four years earlier had revealed him as a boy with violent instincts, a fact that had at least partly determined the choice of both rugby and medical study for him; the demanding physical contest and the practice of surgery, it was thought, would help to channel his aggressive tendencies. Whenever a boy shrieked in pain, Canessa screamed at him to shut up, although he knew the boy could not be blamed. It was in this way that the idea entered his inventive mind of building some kind of hammock in which the most severely injured could sleep at night, away from their companions.

When he suggested it the next morning, his idea was met with contempt. "You're a fool," they said. "You'll kill us all with your hammock."

"Well, at least let me try," said Canessa, and with Daniel Maspons he began to look around for suitable materials. The Fairchild had been designed so that the seats could be removed and the passenger compartment used for freight. For this purpose a large number of nylon straps and poles had been stowed in the luggage area. They were fitted with nozzles which clipped into various fittings in the passenger cabin. Canessa and Maspons discovered that if they took two of the metal poles with the webbing between them, placed them in the joint of the floor and the wall on the left-hand side of the plane, put straps on the other end and attached the straps to the fittings on the top of the ceiling, it created a hammock which swung horizontally away from the floor because of the tilt of the plane. The poles did not stay parallel, but the hammock was large enough for two of the wounded to sleep undisturbed.

They also found that the door which had stood between the passenger compartment and the luggage area could be suspended in the same way and that a seat could be swung up to make a bunk on which two others could sleep. That night Platero lay on the door. Two of the boys with broken legs went up on the bunk, and another, with one of his friends, slept on the hammock. They were all more comfortable, and the others were spared their repeated cries of agony—but in solving one problem they had created another. The warmth of other bodies had been lost and, being suspended fully in the path of the freezing wind which blew into the plane past the inadequate barricade, they suffered acutely from the cold. They were given extra blankets, but this did not make up for the lost warmth from their friends' bodies. The choice had to be made between the pain of the biting cold and the agony of lying with the others.

8

By Monday morning, the fourth day, some of the most seriously injured had started to show signs of recovery in spite of the rudimentary medical attention. Many were still in considerable pain, but much of the swelling had gone down and open wounds had started to heal.

Vizintín, whom Canessa and Zerbino had thought might die from loss of blood, would now call Zerbino to help him out to urinate into the snow. His urine was a dark brown color, which led Zerbino to tell his patient that he might have hepatitis.

"That's all I need," said Vizintín, before stumbling back to his berth in the luggage compartment.

Parrado's recovery was remarkably rapid despite the strain of nursing Susana. The effect of her condition was not to make him despair. Quite the contrary, as he regained his strength there grew within him a lunatic determination to escape. While most of his companions only thought of being rescued, Parrado considered the active option of somehow getting back to civilization by his own efforts, and he confided this determination to Carlitos Páez, who also wanted to leave.

"Impossible," said Carlitos. "You'd freeze to death in the snow."

"Not if I wore enough clothes."

"Then you'd starve to death. You can't climb mountains on a little piece of chocolate and a sip of wine."

"Then I'll cut meat from one of the pilots," said Parrado. "After all, they got us into this mess."

Carlitos was not shocked by this because he did not take it seriously. He was, however, among those who were increasingly concerned at the length of time it was taking to rescue them. It was now four days since the plane had crashed, and apart from the biplane which had dipped its wings while flying over them the day before, there had been no sign that the outside world even knew that any of them was still alive. Because the thought that they could not be seen from the air, or that they had been given up for lost, was so terrible, few of the survivors would permit it to enter their minds. They evolved the theory that they had been seen but were too high in the cordillera to be rescued by helicopter, so an expedition was coming by land. Marcelo believed this, and so did Pancho Delgado, a law student who hobbled around the plane on his one good leg talking cheerfully to the others and eloquently insisting that God would not forsake them in their predicament.

Most of the boys were grateful to Delgado because he calmed the panic that was rising within them. They were less favorably disposed toward the smaller group of pessimists—notably Canessa, Zerbino, Parrado, and the Strauch cousins—who questioned the hypothesis that help was on its way.

"Why," asked Fito Strauch, "if they know where we are, don't they drop some supplies?"

"Because they know that they'd be buried in the snow," said Marcelo, "and we wouldn't be able to get at them."

None of the boys had any real idea of where they were. They had found charts in the pilots' cabin which they studied for hour after hour, huddled out of the wind in the dark cabin. None of them knew how to read aeronautical charts, but Arturo Nogueira, a shy, withdrawn boy who had broken both his legs, appointed himself map reader to the group and found Curicó among the many towns and villages. They remembered that the copilot had said repeatedly that they had passed Curicó, and it was plain from the map that Curicó was well into Chile on the western side of the Andes. Thus they must be somewhere in the foothills. The needle on the altimeter pointed to 7,000 feet. To the west, Chilean villages could not be very far away.

The difficulty facing them was that any path to the west was blocked by the gigantic mountains, and the valley in which they were trapped led to the east—back, they thought, into the middle of the cordillera. They were convinced that if only they could climb to the top of the mountain to the west, they would be met by a view of green valleys and Chilean farmhouses.

They had been able to walk away from the plane only until nine or so in the morning. After that, if there was any sun, the crust soon melted and they sank up to their thighs in the soft powder. This had so far prevented any of them from venturing more than a few yards from the plane, for they feared that, like Valeta, they might simply disappear in the deep snow. Fito Strauch, their inventor, discovered, however, that if cushions from the passenger seats were tied to their boots, they made passable snowshoes. Walking in this way was difficult but it was possible, and both he and Canessa immediately wanted to set off up the mountain, not only to see what was on the other side but to discover if any of their friends—and in Fito's case, another of his cousins—had survived the crash and were living in the tail.

There were other incentives. Roque had told them that in the tail they would find batteries to power the VHF radio. There might also be suitcases scattered down the mountainside, for the track the plane had followed was still visible in the snow. These would provide them with extra clothes.

Carlitos Páez and Numa Turcatti—among others—were also eager to climb the mountain, and at seven o'clock on the morning of Tuesday, October 17, all four set out. The sky was clear but it was still bitterly cold so the surface was firm. Wearing rugby boots, they made quite good progress. Canessa wore some gloves he had made out of a pair of socks.

They walked for an hour, rested, and then walked on. The air was thin and the going hard; as the sun rose in the sky, the crust melted and they had to tie on the cushions, which soon became sodden. To avoid one foot stepping on the other, they had to walk bowlegged.

None of them had eaten anything substantial for nearly five days, and soon Canessa suggested that they turn back. He was overruled and they struggled on, but a short time later Fito sank waist deep in snow on the brink of a crevasse. This frightened them all. The plane beneath them seemed small in the vast landscape; the boys around it were specks in the snow. There were no suitcases to be seen and no sign of the tail.

"It's not going to be easy getting out of here," said Canessa.

"But if we aren't rescued, we'll have to walk out," said Fito.

"We'd never make it," said Canessa. "Look how weak we've become without food."

"Do you know what Nando said to me?" Carlitos said to Fito. "He said that if we weren't rescued, he'd eat one of the pilots to get out of here." There was a pause; then Carlitos added, "That hit on the head must have made him slightly mad."

"I don't know," said Fito, his honest, serious features quite composed. "It might be the only way to survive."

Carlitos said nothing, and they turned to go back down the mountain.

9

The experience of the expedition depressed them all. During the following days Liliana Methol continued to comfort those who were afraid, and Pancho Delgado did his best to keep up their spirit of optimism, but the days were passing without the least sign that rescue might come, and they had all seen how the toughest among them had fared on a short climb up the side of the mountain. What hope, then, could there be for the weak and injured?

They piled into the plane each night and lay in the freezing darkness with their thoughts of home and family. Gradually they dropped off to sleep, the feet of one on the shoulders of another, Javier and Liliana together, Echavarren and Nogueira on the hammock, Nando and Susana Parrado lying in each other's arms.

On their eighth night on the mountain, Parrado awoke and felt that Susana had grown cold and still in his embrace. The warmth and the movement of breathing—both were gone. At once he pressed his mouth to hers and with tears streaming down his cheeks blew air into her lungs. The other boys awoke and watched and prayed as Parrado tried to revive his sister. When exhaustion forced him to give up, Carlitos Páez took over the task, but it was all to no avail. Susana was dead.

Two

1

When Air Traffic Control at the airport of Pudahuel in Santiago first lost contact with the Uruguayan Fairchild on the afternoon of Friday, October 13, they immediately telephoned the Servicio Aéreo de Rescate (the Aerial Rescue Service), with headquarters at Santiago's other airport, Los Cerrillos. The commander of the SAR was away, so two former commanders were called in to direct the search-and-rescue operation—Carlos García and Jorge Massa. They were Chilean Air Force officers who were trained not just to command but to fly all the types of aircraft they had at their disposal: Douglas C-47s, DC-6s, Twin Otters and Cessna light aircraft, and the powerful Bell helicopters.

That afternoon a DC-6 began to search on a path starting from the last reported position of the missing aircraft, the air corridor from Curicó to Angostura and Santiago. The populated areas in the zone were ignored because any crash would have been reported; the search was focused on the more mountainous areas. Finding nothing, they moved farther back along the supposed route of the Fairchild to the area between Curicó and

Planchon. There was a snowstorm over Planchon itself, so nothing could be seen, and the DC-6 returned to Santiago.

The next day García and Massa analyzed more precisely the information they had at hand: the time the Fairchild left Mendoza, the time it flew over Malargüe, the speed of the plane, and the head wind it faced in flying over the Andes. They concluded that the plane could not possibly have been over Curicó when the pilot had reported this position, but over Planchon, so that instead of turning toward Angostura and Santiago and descending to the airport of Pudahuel, the Fairchild had turned into the middle of the Andes and flown down into the area of the Tinguiririca, Sosneado, and Palomo mountains. With great precision García and Massa plotted a twenty-inch square on their map, representing the area in which the crash must have occurred. They then sent planes out from Santiago to cover it.

The difficulties presented were obvious. The mountains there rose to 15,000 feet. If the Fairchild had crashed anywhere among them, it would certainly have fallen into one of the valleys which lay around 12,000 feet and were covered with twenty to a hundred feet of white snow. Since the Fairchild had a white roof, it would be virtually invisible to an airplane flying above the level of the peaks. To fly in the turbulence between the peaks was a sure way to lose further planes and lives, but a methodical search of the whole area was a ritual they were obliged to perform.

From the very start the professionals in the control room of the SAR at the airport of Los Cerrillos had little hope that anyone could have survived a crash in the middle of the cordillera. They knew that the temperature at that altitude at that time of year went down to 30 or 40 degrees below zero, so that if, by some quirk of fate, a few of the passengers had survived the crash, they would certainly have died of cold during their first night on the mountain.

There is, however, an international convention that the country in which an accident occurs will search for the wreck for ten days, and in spite of the political and economic chaos that Chile faced at the time, it was a duty the SAR had to perform. Moreover, the relatives of the passengers had started to arrive in Santiago.

2

For those at home, the hours immediately following the first reports of the plane's disappearance had been filled with confusion as well as desperate anxiety. After the initial broadcasts which told that the Fairchild had stopped at Mendoza—which none of the parents had known—and taken off the next day and disappeared, official silence set in and a number of conflicting reports from unofficial sources rushed into the vacuum. Daniel Fernández's father learned on Saturday the fourteenth that the plane had been "found" even before he had known that it was missing, for he had not listened to the radio the night before. Others heard that the boys had arrived safely and were booked into their hotel in Santiago, and there was still another rumor that they had landed not in Santiago but somewhere in the south of Chile.

The means through which some of the reports were first conveyed and subsequently corrected was a radio in a Carrasco home. Rafael Ponce de León was a radio ham, a hobby he had inherited from his father, who had installed a whole range of sophisticated equipment—including a powerful transmitter, a Collins KWM2—in the basement of their house. Rafael was also an Old Christian and a friend of Marcelo Pérez. He himself had not joined the trip to Santiago only because he was reluctant to leave his wife, who was seven months pregnant. At Marcelo's request, Rafael had used his radio to book rooms for the rugby team in Santiago, calling a fellow ham in Chile who had connected him to the Chilean telephone system. Quicker and cheaper than telephoning, the practice was not strictly legal but it was tolerated.

When he heard late on the thirteenth that the Fairchild was lost in the Andes, he went at once to his radio. He contacted the Crillon Hotel in Santiago and was told that the Old Christians had checked in. When later reports cast this in doubt, Rafael called the hotel again and discovered that only two of the players were there, two who had flown to Santiago on scheduled flights, one—Gilberto Regules—because he had missed the Fairchild; the other, Bobby Jaugust, because his father represented KLM in Montevideo.

No sooner had he scotched this rumor than the telephone rang with the news that the Chilean *novia* of one of the boys on the plane had spoken with her future parents-in-law, the Magris, to say that the plane had landed in a small town in southern Chile and all on board were safe. Having raised false hopes with the rumor of the Hotel Crillon, Rafael was determined to check this one, and he got in touch with the Uruguayan chargé d'affaires in Santiago, César Charlone. Charlone told him that he thought it unlikely; the official news was still that the plane was lost.

By now the story that the plane had been found had spread from the Magris to half the families involved. Fearing that their hopes might be false, Rafael decided to go to the source of the rumor, the *novia* of Guido Magri, María de los Angeles. He contacted her through his radio and asked her if it was true. María confessed that it was not. Señora Magri had sounded so depressed over the telephone that she had told her this white lie. "I was so sure it would be found," she said, "that I told her it had been found already."

Rafael recorded what she had said and later that night sent the tape to Radio Monte Carlo for use in their next news bulletin. It was well past midnight when he closed down his transmitter, but his work had not been in vain. By nine that morning the rumor was dead. The plane had not been found.

Carlos Páez Vilaró, a well-known painter and the father of Carlitos, was the first to arrive at SAR headquarters at Los Cerrillos. He had heard the news of the plane's disappearance at his former wife's home in Carrasco—only by chance, as he left one of his daughters there Friday afternoon, for since the divorce the children had lived with their mother, Madelón Rodríguez. He had made what inquiries he could in official circles, from the Uruguayan chargé d'affaires in Santiago and a Uruguayan Air Force officer whom he knew personally. Charlone, the chargé d'affaires, had not been reassuring, and although the Air Force officer called Farradas "the best and most experienced" pilot in the force, Páez Vilaró knew that his friend and Farradas were the only two surviving pilots of their generation; the rest had been killed in accidents. Telling Madelón that he himself would find the boys, he set out for Santiago early Saturday morning.

That afternoon he flew in an Air Force DC-6 along the likely route of the Fairchild. When he returned to the airport another of the boys' relatives had arrived, and by the next day there were a total of twenty-two on the scene.

Faced with this flood, Commander Massa announced that no more relatives would be permitted to fly on the planes taking part in the search, so they congregated instead in the office of César Charlone. There they heard the news that a miner named Camilo Figueroa had told the Chilean police that he had seen the Fairchild fall in flames about seventy miles northeast of Curicó in the zone of El Tiburcio.

On Monday the sixteenth García and Massa directed the search in that area. Nothing was seen in the morning, but in the afternoon a pilot reported from El Tiburcio that smoke could be seen rising from the mountains. Closer inspection revealed that the smoke came from the hut of a hill farmer.

On the same day search parties set off overland made up of *carabineros* (Chilean militarized police) and members of the Cuerpo de Socorro Andino, a body of volunteers formed to rescue those who are lost in the Andes. They left from Rancagua and made for the area between Planchon and El Tiburcio but were stopped in the afternoon by heavy snow and strong wind.

Those weather conditions also grounded the planes the next day and the day after that, October 17 and 18. Impenetrable clouds and snow covered the whole area of thc search. Dispirited, some of the relatives returned to Montevideo. Others remained and began to think of mounting a search of their own. It was not that they felt that the Chileans were doing less than was possible—even the sister craft of the Fairchild which had been sent to assist the Chileans by the Uruguayan Air Force had been grounded by the weather—but they knew that time was running out and that the professionals had no real faith that their sons might be alive. "Impossible," Commander Massa had said to the press. "And if anyone were alive he would sink in the snow."

Páez Vilaró considered what he might do on his own. He found a book in a shop in Santiago entitled *The Snows and Mountains of Chile* which contained the information that the land covered by the Tinguiririca and Palomo mountains belonged to a certain Joaquín Gandarillas. Reasoning that the man who owned the land must know it best, Páez Vilaró went to see Gandarillas, who received him politely but explained that his huge estate had recently been confiscated under President Allende's program for agrarian reform. Nevertheless, Gandarillas knew the land like the back of his hand, and by the time their meeting was over, Páez Vilaró had persuaded him to leave with him the next day for the area of the Tinguiririca volcano.

A two-day trip by car and on horseback brought them to the western slope of the mountain. The heavy snow had ceased but the freshness of the fall only emphasized the emptiness of the place. There was nothing to be seen—either living or dead—and yet Páez Vilaró stood staring at the immense mass of the mountain and whistled, thinking that by magic the sound might reach his son. The whistle echoed in the rock and was muffled by the snow. There was nothing to do but return.

While Páez Vilaró was taking these steps to find his son, there were some at home who had recourse to less orthodox methods of search and rescue—among them the mother of his former wife, Madelón.

Accompanied by Javier Methol's brother, Juan José, on October 16 she went to see an old man in Montevideo, a professional water diviner who was said to have powers of clairvoyance that could seek out far more than hidden springs. With them they took a map of the Andes. When the old man held out his forked stick over the map, the stick quivered and then fell at a point on the map on the eastern slope of the Tinguiririca volcano nineteen miles from the spa of Termas del Flaco.

Her daughter Madelón reported this position to Páez Vilaró in Chile, through Rafael Ponce de León's radio, but Páez Vilaró told her that the SAR had already made a thorough search in that area and that, if they had crashed there, there would be no chance they had survived.

This last was something that Madelón never wished to hear, so she put the water diviner out of her mind. Yet the idea of help from a clairvoyant stayed with her. She went to the Uruguayan astrologer Boris Cristoff and asked him for the name of the best clairvoyant in the world.

"Croiset," he said without hesitation. "Gerard Croiset in Utrecht."

Rosina Strauch, Fito's mother, hoped for help from another source. The Virgin of Garabandal, she had been told, had appeared to some children in Spain about ten years earlier, an apparition never accepted by the Vatican. To persuade the Pope of it, the Virgin must surely want to perform a miracle. If so, here was her chance, and in this belief Rosina and two other mothers began to pray to the Virgin of Garabandal.

But others had already resigned themselves to their loss, and their prayers were for the strength to bear it and for their sons' souls. For the mother of Carlos Valeta—the boy who had disappeared in the snow right after the crash—hope was impossible. On Friday afternoon she had a vision, first of a falling plane, then of her son's wounded face, then of him sleeping, and by five thirty she had known he was dead. Other parents' reasons for resignation were based on the conclusion that they could not escape, that surviving a crash for several days in the cordillera was out of the question.

Nevertheless, Rafael Ponce de León's basement each evening was crowded with relatives, friends, and girlfriends of the boys, desperate for news.

The search by the SAR was resumed on October 19. It continued throughout that day and the next and into the morning of the twenty-first. At the same time, sorties were flown by Argentinian planes from Mendoza. Páez Vilaró and others continued to search in a Cessna lent to them by the Aero Club of San Fernando. In spite of all this effort, there was no trace of the Fairchild.

The search had been going on for eight days, two of which were lost because of the weather. The lives of SAR men were being risked, and expensive fuel was being burned up, in a search which all reasonable men must know was futile. Thus at midday on the twenty-first, commanders García and Massa announced that "the search for the Uruguayan Aircraft 571 is canceled because of negative results."

Three

1

On the morning of the ninth day, the body of Susana Parrado was dragged out onto the snow. No sound but the wind met the ears of the survivors as they stumbled from the cabin; nothing was to be seen but the same monotonous arena of rock and snow.

As the light changed, the mountains took on different moods and appearances. Early in the morning, they seemed bright and distant. Then, as the day progressed, shadows lengthened and the gray, reddish, and green stone became the features of brooding beasts or disgruntled gods frowning down upon the intruders.

The seats of the plane were laid out on the snow like deck chairs on the veranda of an *estancia*. Here, the first out would sit down to melt snow for drinking water while staring at the horizon. Each could see in the face of his companions the rapid progress of their physical deterioration. The movements of those who busied themselves in the cabin or around the fuselage had grown heavy and slow. They were all exhausted by the slightest exertion. Many remained sitting where they had slept, too listless and depressed even to go out into the fresh air. Irritability was an increasing problem.

Marcelo Pérez, Daniel Fernández, and the older members of the group feared that some of the boys were on the verge of hysteria. The waiting was wearing them down. They had started to squabble among themselves.

Marcelo did what he could to set an example. He was optimistic and he was fair. He talked confidently of rescue and tried to get his team to sing songs. There was one desultory rendering of "Clementine," but no one had the spirit to sing. It was also becoming evident to them all that their captain was not as confident as he seemed. At night he was overtaken by melancholy; his mind turned to his mother and how much she must be suffering, to his brother on a honeymoon in Brazil, and to the rest of his family. He tried to hide his sobs from the others, but if he slept he would dream and wake screaming. His friend Eduardo Strauch did his best to comfort him, but Marcelo felt that as captain of the team—and chief exponent of the trip to Chile—he had been responsible for what had happened.

"Don't be a fool," said Eduardo. "You can't look at things like that. I persuaded Gaston and Daniel Shaw to come and they're both dead. I even rang Daniel to remind him, but I don't feel responsible for his death."

"If anyone's responsible," said his cousin Fito, "it's God. Why did He let Gaston die?" Fito was referring to the fact that Gaston Costemalle, who had fallen out of the back of the plane, was not the first of his family to die; his mother had already lost her husband and other son. "Why does God let us suffer like this? What have we done?"

"It's not as simple as that," said Daniel Fernández, the third of the Strauch cousins.

There were two or three among the twenty-seven whose courage and example acted as pillars to their morale. Echavarren, in considerable pain from his smashed leg, remained cheerful and outgoing, screaming and cursing at anyone who stepped on him but always making up for it afterward with a courteous apology or a joke. Enrique Platero was energetic and brave, despite the wound in his stomach. And Gustavo Nicolich made his "gang" get up in the morning, tidy the cabin, and then play such games as charades, while at night he persuaded them to say the rosary with Carlitos Páez.

Liliana Methol, the one woman among them, was a unique source of solace. Though younger, at thirty-five, than their mothers, she became for them all an object of filial affection. Gustavo Zerbino, who was only nineteen, called her his godmother, and she responded to him and to the others with comforting words and gentle optimism. She too realized that the boys' morale was in danger of collapse and thought of ways to distract them from their predicament. On the evening of that ninth day, she gathered them around her and suggested that each tell an anecdote from his past life. Few of them could think of anything to say. Then Pancho Delgado volunteered to tell three stories, all about his future father-in-law.

When he had first met his *novia*, he told them, she was only fifteen, while he was three or four years older. He was not at all sure whether her parents would welcome him, and he was anxious about the impression he would create. Within a short time, Delgado reported, he had accidentally pushed her father into a swimming pool, injuring his leg; he had discharged a shotgun into the roof of their family car, a brand-new BMW 2002, leaving an enormous hole with pieces of metal bent back like the petals of a flower; and he had very nearly electrocuted her father while helping him prepare for a party in the garden of their house in Carrasco.

His anecdotes were like a tonic to the boys sitting in the dank atmosphere of the plane, waiting to feel tired enough to sleep, and for this they felt grateful. Yet when the turn came for other stories, no one spoke, and as the light faded, each returned to his own thoughts.

2

They awoke on the morning of Sunday, October 22, to face their tenth day on the mountain. First to leave the plane were Marcelo Pérez and Roy Harley. Roy had found a transistor radio between two seats and by using a modest knowledge of electronics, acquired when helping a friend construct a hi-fi system, he had been able to make it work. It was difficult to receive signals in the deep cleft between the huge mountains, so Roy made an aerial with strands of wire from the plane's electric circuits. While he turned the dial, Marcelo held the aerial and moved it around. They picked up scraps of broadcasts from Chile but no news of the rescue effort. All that came over the radio waves were the strident voices of Chilean politicians embroiled in the strike by the middle classes against the socialist government of President Allende.

Few of the other boys came out into the snow. Starvation was taking its effect. They were becoming weaker and more listless. When they stood up they felt faint and found it difficult to keep their balance. They felt cold, even when the sun rose to warm them, and their skin started to grow wrinkled like that of old men.

Their food supplies were running out. The daily ration of a scrap of chocolate, a capful of wine, and a teaspoonful of jam or canned fish—eaten slowly to make it last—was more torture than sustenance for these healthy, athletic boys; yet the strong shared it with the weak, the healthy with the injured. It was clear to them all that they could not survive much longer. It was not so much that they were consumed with ravenous hunger as that they felt themselves grow weaker each day, and no knowledge of medicine or nutrition was required to predict how it would end.

Their minds turned to other sources of food. It seemed impossible that there should be nothing whatsoever growing in the Andes, for even the meanest form of plant life might provide some nutrition. In the immediate vicinity of the plane there was only snow. The nearest soil was a hundred feet beneath them. The only ground exposed to sun and air was barren mountain rock on which they found nothing but brittle lichens. They scraped some of it off and mixed it into a paste with melted snow, but the taste was bitter and disgusting, and as food it was worthless. Except for lichens there was nothing. Some thought of the cushions, but even these were not stuffed with straw. Nylon and foam rubber would not help them.

For some days several of the boys had realized that if they were to survive they would have to eat the bodies of those who had died in the crash. It was a ghastly prospect. The corpses lay around the plane in the snow, preserved by the intense cold in the state in which they had died. While the thought of cutting flesh from those who had been their friends was deeply repugnant to them all, a lucid appreciation of their predicament led them to consider it.

Gradually the discussion spread as these boys cautiously mentioned it to their friends or to those they thought would be sympathetic. Finally, Canessa brought it out into the open. He argued forcefully that they were not going to be rescued; that they would have to escape themselves, but that nothing could be done without food; and that the only food was human flesh. He used his knowledge of medicine to describe, in his penetrating, high-pitched voice, how their bodies were using up their reserves. "Every time you move," he said, "you use up part of your own body. Soon we shall be so weak that we won't have the strength even to cut the meat that is lying there before our eyes."

Canessa did not argue just from expediency. He insisted that they had a moral duty to stay alive by any means at their disposal, and because Canessa was earnest about his religious belief, great weight was given to what he said by the more pious among the survivors.

"It is meat," he said. "That's all it is. The souls have left their bodies and are in heaven with God. All that is left here are the carcasses, which are no more human beings than the dead flesh of the cattle we eat at home."

Others joined the discussion. "Didn't you see," said Fito Strauch, "how much energy we needed just to climb a few hundred feet up the mountain? Think how much more we'll need to climb to the top and then down the other side. It can't be done on a sip of wine and a scrap of chocolate."

The truth of what he said was incontestable.

A meeting was called inside the Fairchild, and for the first time all twenty-seven survivors discussed the issue which faced them—whether or not they should eat the bodies of the dead to survive. Canessa, Zerbino, Fernández, and Fito Strauch repeated the arguments they had used before. If they did not they would die. It was their moral obligation to live, for their own sake and for the sake of their families. God wanted them to live, and He had given them the means to do so in the dead bodies of their friends. If God had not wished them to live, they would have been killed in the accident; it would be wrong now to reject this gift of life because they were too squeamish.

"But what have we done," asked Marcelo, "that God now asks us to eat the bodies of our dead friends?"

There was a moment's hesitation. Then Zerbino turned to his captain and said, "But what do you think *they* would have thought?"

Marcelo did not answer.

"I know," Zerbino went on, "that if my dead body could help you to stay alive, then I'd certainly want you to use it. In fact, if I do die, and you don't eat me, then

I'll come back from wherever I am and give you a good kick in the ass."

This argument allayed many doubts, for however reluctant each boy might be to eat the flesh of a friend, all of them agreed with Zerbino. There and then they made a pact that if any more of them were to die, their bodies were to be used as food.

Marcelo still shrank from a decision. He and his diminishing party of optimists held onto the hope of rescue, but few of the others any longer shared their faith. Indeed, a few of the younger boys went over to the pessimists—or the realists, as they considered themselves—with some resentment against Marcelo Pérez and Pancho Delgado. They felt they had been deceived. The rescue they had been promised had not come.

The latter were not without support, however. Coche Inciarte and Numa Turcatti, both strong, tough boys with an inner gentleness, told their companions that while they did not think it would be wrong, they knew that they themselves could not do it. Liliana Methol agreed with them. Her manner was calm as always but, like the others, she grappled with the emotions the issue aroused. Her instinct to survive was strong, her longing for her children was acute, but the thought of eating human flesh horrified her. She did not think it wrong; she could distinguish between sin and physical revulsion, and a social taboo was not a law of God. "But," she said, "as long as there is a chance of rescue, as long as there is *something* left to eat, even if it is only a morsel of chocolate, then I can't do it."

Javier Methol agreed with his wife but would not deter others from doing what they felt must be done. No one suggested that God might want them to choose to die. They all believed that virtue lay in survival and that eating their dead friends would in no way endanger their souls, but it was one thing to decide and another to act.

Their discussions had continued most of the day, and by midafternoon they knew that they must act now or not at all, yet they sat inside the plane in total silence. At last a group of four—Canessa, Maspons, Zerbino, and Fito Strauch—rose and went out into the snow. Few followed them. No one wished to know who was going to cut the meat or from which body it was to be taken.

Most of the bodies were covered by snow, but the buttocks of one protruded a few yards from the plane. With no exchange of words, Canessa knelt, bared the skin, and cut into the flesh with a piece of broken glass. It was frozen hard and difficult to cut, but he persisted until he had cut away twenty slivers the size of matchsticks. He then stood up, went back to the plane, and placed them on the roof.

Inside there was silence. The boys cowered in the Fairchild. Canessa told them that the meat was there on the roof, drying in the sun, and that those who wished to do so should come out and eat it. No one came, and again Canessa took it upon himself to prove his resolution. He prayed to God to help him do what he knew to be right and then took a piece of meat in his hand. He hesitated. Even with his mind so firmly made up, the horror of the act paralyzed him. His hand would neither rise to his mouth nor fall to his side while the revulsion which possessed him struggled with his stubborn will. The will prevailed. The hand rose and pushed the meat into his mouth. He swallowed it.

He felt triumphant. His conscience had overcome a primitive, irrational taboo. He was going to survive.

Later that evening, small groups of boys came out of the plane to follow his example. Zerbino took a strip and swallowed it as Canessa had done, but it stuck in his throat. He scooped a handful of snow into his mouth and managed to wash it down. Fito Strauch followed his example, then Maspons and Vizintín and others.

Meanwhile Gustavo Nicolich, the tall, curly-haired boy, only twenty years old, who had done so much to keep up the morale of his young friends, wrote to his *novia* in Montevideo.

> *Most dear Rosina:*
>
> *I am writing to you from inside the plane (our* petit hotel *for the moment). It is sunset and has started to be rather cold and windy which it usually does at this hour of the evening. Today the weather was wonderful—a beautiful sun and very hot. It reminded me of the days on the beach with you—the big difference being that then we would be going to have lunch at your place at midday whereas now I'm stuck outside the plane without any food at all.*
>
> *Today, on top of everything else, it was rather depressing and a lot of the others began to get discouraged (today is the tenth day we have been here), but luckily this gloom did not spread to me because I get incredible strength just by thinking that I'm going to see you again. Another of the*

things leading to the general depression is that in a while the food will run out: we have only got two cans of seafood (small), one bottle of white wine, and a little cherry brandy left, which for twenty-six men (well, there are also boys who want to be men) is nothing.

One thing which will seem incredible to you—it seems unbelievable to me—is that today we started to cut up the dead in order to eat them. There is nothing else to do. I prayed to God from the bottom of my heart that this day would never come, but it has and we have to face it with courage and faith. Faith, because I came to the conclusion that the bodies are there because God put them there and, since the only thing that matters is the soul, I don't have to feel great remorse; and if the day came and I could save someone with my body, I would gladly do it.

I don't know how you, Mama, Papa, or the children can be feeling; you don't know how sad it makes me to think that you are suffering, and I constantly ask God to reassure you and give us courage because that is the only way of getting out of this. I think that soon there will be a happy ending for everyone.

You'll get a shock when you see me. I am dirty, with a beard, and a little thinner, and with a big gash on my head, another one on my chest which has healed now, and one very small cut which I got today working in the cabin of the plane, besides various small cuts in the legs and on the shoulder; but in spite of it all, I'm all right.

3

Those who first peered through the portholes of the plane the next morning could see that the sky was overcast but that a little sun shone through the clouds onto the snow. Some darted cautious looks toward Canessa, Zerbino, Maspons, Vizintín, and the Strauch cousins. It was not that they thought that God would have struck them down, but they knew from their *estancias* that one should never eat a steer that dies from natural causes, and they wondered if it might not be just as unhealthy to do the same with a man.

The ones who had eaten the meat were quite well. None of them had eaten very much and in fact they felt as enfeebled as the others. As always, Marcelo Pérez was the first to raise himself from the cushions.

"Come on," he said to Roy Harley. "We must set up the radio."

"It's so cold," said Roy. "Can't you get someone else?"

"No," said Marcelo. "It's your job. Come on."

Reluctantly Roy took his shoes down from the hat rack and put them on over his two pairs of socks. He squeezed himself out of the line of dozing figures and climbed over those nearest the entrance to follow Marcelo out of the plane. One or two others followed him out.

Marcelo had already taken hold of the aerial and was waiting while Roy picked up the radio, switched it on, and began to turn the dial. He tuned it to a station in Chile which the day before had broadcast nothing but political propaganda; now, however, as he held the radio to his ear, he heard the last words of a news bulletin. "The SAR has requested all commercial and military aircraft overflying the cordillera to check for any sign of the wreckage of the Fairchild Number Five-seventy-one. This follows the cancellation of the search by the SAR for the Uruguayan aircraft because of negative results."

The newscaster moved on to a different topic. Roy took the radio away from his ear. He looked up at Marcelo and told him what he had heard. Marcelo dropped the aerial, covered his face with his hands, and wept with despair. The others who had clustered around Roy, upon hearing the news, began to sob and pray, all except Parrado, who looked calmly up at the mountains which rose to the west.

Gustavo Nicolich came out of the plane and, seeing their faces, knew what they had heard.

"What shall we tell the others?" he asked.

"We mustn't tell them," said Marcelo. "At least let them go on hoping."

"No," said Nicolich. "We must tell them. They must know the worst."

"I can't, I can't," said Marcelo, still sobbing into his hands.

"I'll tell them," said Nicolich, and he turned back toward the entrance to the plane.

He climbed through the hole in the wall of suitcases and rugby shirts, crouched at the mouth of the dim tunnel, and looked at the mournful faces which were turned toward him.

"Hey, boys," he shouted, "there's some good news! We just heard it on the radio. They've called off the search."

Inside the crowded cabin there was silence. As the hopelessness of their predicament enveloped them, they wept.

"Why the hell is that good news?" Páez shouted angrily at Nicolich.

"Because it means," he said, "that we're going to get out of here on our own."

The courage of this one boy prevented a flood of total despair, but some of the optimists who had counted on rescue were unable to rally. The pessimists, several of them as unhopeful about escape as they had been about rescue, were not shocked; it was what they had expected. But the news broke Marcelo. His role as their leader became empty and automatic, and the life went out of his eyes. Delgado, too, was changed by the news. His eloquent and cheerful optimism evaporated into the thin air of the cordillera. He seemed to have no faith that they would get out by their own efforts and quietly withdrew into the background. Of the old optimists, only Liliana Methol still offered hope and consolation. "Don't worry," she said. "We'll get out of here, all right. They'll find us when the snow melts." Then, as if remembering how little food remained besides the bodies of the dead, she added, "Or we'll walk to the west."

To escape: that was the obsession of the new optimists. It was disconcerting that the valley in which they were trapped ran east, and that to the west there was a solid wall of towering mountains, but this did not deter Parrado. No sooner had he learned of the cancellation of the search than he announced his intention of setting off—on his own, if necessary—to the west. It was only with great difficulty that the others restrained him. Ten days before he had been given up for dead. If anyone was going to climb the mountains, there were others in a much better physical condition to do so. "We must think this out calmly," said Marcelo, "and act together. It's the only way we'll survive."

There was still sufficient respect for Marcelo and enough team discipline in Parrado to accept what the others decided. He was not alone, however, in his insistence that, before they got any weaker, another expedition should set out, either to climb the mountain and see what was on the other side or to find the tail.

It was agreed that a group of the fittest among them should set off at once, and a little more than an hour after they had heard the news on the radio, Zerbino, Turcatti, and Maspons set off up the mountain, watched by their friends.

Canessa and Fito Strauch returned to the corpse they had opened the day before and cut more meat off the bone. The strips they had put on the roof of the plane had now all been eaten. Not only were they easier to swallow when dried in the outside air, but the knowledge that they were not going to be rescued had persuaded many of those who had hesitated the day before. For the first time, Parrado ate human flesh. So, too, did Daniel Fernández, though not without the greatest effort of will to overcome his revulsion. One by one, they forced themselves to take and swallow the flesh of their friends. To some, it was merely an unpleasant necessity; to others, it was a conflict of conscience with reason.

Some could not do it: Liliana and Javier Methol,

Coche Inciarte, Pancho Delgado. Marcelo Pérez, having made up his mind that he would take this step, used what authority he still possessed to persuade others to do so, but nothing he said had the effect of a short statement from Pedro Algorta. He was one of the two boys who had been dressed more scruffily at the airport than the others, as if to show that he despised their bourgeois values. In the crash, he had been hit on the head and suffered total amnesia about what had happened the day before. Algorta watched Canessa and Fito Strauch cutting the meat but said nothing until it came to the moment when he was offered a slice of flesh. He took it and swallowed it and then said, "It's like Holy Communion. When Christ died he gave his body to us so that we could have spiritual life. My friend has given us his body so that we can have physical life."

It was with this thought that Coche Inciarte and Pancho Delgado first swallowed their share, and Marcelo grasped it as a concept which would persuade others to follow his example and survive. One by one they did so until only Liliana and Javier Methol remained.

Now that it was established that they were to live off the dead, a group of stronger boys was organized to cover the corpses with snow, while those who were weaker or injured sat on the seats, holding the aluminum water makers toward the sun, catching the drops of water in empty wine bottles. Others tidied the cabin. Canessa, when he had cut enough meat for their immediate needs, made a tour of inspection of the wounded. He was moderately content with what he saw. Almost all the superficial wounds were continuing to heal, and none showed signs of infection. The swelling around broken bones also was subsiding; Alvaro Mangino and Pancho Delgado, for example, both managed, despite considerable pain, to hobble around outside the plane. Arturo Nogueira was worse off; if he went outside the plane he had to crawl, pulling himself forward with his arms. The state of Rafael Echavarren's leg was growing serious; it showed the first indications of gangrene.

Enrique Platero, the boy who had had the tube of steel removed from his stomach, told Canessa that he was feeling perfectly well but that a piece of his insides still protruded from the wound. The doctor carefully unwound the rugby shirt which Platero continued to use as a bandage and confirmed the patient's observation; the wound was healing well but something stuck out from the skin. Part of this projection had gone dry, and Canessa suggested to Platero that if he cut off the dead matter the rest might be more easily pushed back under the skin.

"But what is sticking out?" asked Platero.

Canessa shrugged his shoulders. "I don't know," he said. "It's probably part of the lining of the stomach, but if it's the intestine and I cut it open, you've had it. You'll get peritonitis."

Platero did not hesitate. "Do what you have to do," he said, and lay back on the door.

Canessa prepared to operate. As scalpel he had a choice between a piece of broken glass or a razor blade. His sterilizer was the subzero air all around them. He disinfected the area of the wound with eau de cologne and then carefully cut away a small slice of the dead skin with the glass. Platero did not feel it, but the protruding gristle still would not go back under the skin. With even greater caution Canessa now cut yet closer to the living tissue, dreading all the time that he might cut into the intestine, but again he seemed to have done no harm and this time, with a prod from the surgeon's finger, the gut retired into Platero's stomach where it belonged.

"Do you want me to stitch you up?" Canessa asked his patient. "I should warn you that we don't have any surgical thread."

"Don't worry," said Platero, rising on his elbows and looking down at his stomach. "This is fine. Just tie it up again and I'll be on my way."

Canessa retied the rugby shirt as tightly as he could, and Platero swung his legs off the door and got to his feet. "Now I'm ready to go on an expedition," he said, "and when we get back to Montevideo I'll take you on as my doctor. I couldn't possibly hope for a better one."

Outside the plane, following the example of Gustavo Nicolich, Carlitos Páez was writing to his father, his mother, and his sisters. He also wrote to his grandmother:

> *You can have no idea how much I have thought about you because I love you, I adore you, because you have already received so many blows in your life, because I don't know how you are going to stand this one. You, Buba, taught me many things but the most important one was faith in God. That has increased so much now that you cannot conceive of it. . . . I want you to know that you are the kindest grandmother in the world and I shall remember you each moment I am alive.*

4

Zerbino, Turcatti, and Maspons followed the track of the plane up the mountain. Every twenty or twenty-five steps the three were forced to rest, waiting for their hearts to beat normally again. The mountain seemed almost vertical, and they had to clutch at the snow with their bare hands. They had left in such a hurry that they had not thought of how they should equip themselves for the climb. They wore only sneakers or moccasins and shirts, sweaters, and light jackets, with thin trousers covering their legs. All three were strong, for they were players who had been in training, but they had barely eaten for the past eleven days.

The air that afternoon was not so cold. As they climbed, the sun shone on their backs and kept them warm. It was their feet, sodden with freezing snow, which suffered most. In the middle of the afternoon they reached a rock, and Zerbino saw that the snow around it was melting. He threw himself down and sucked at drops of water suspended from the disintegrating crystals. There was also another form of lichen, which he put into his mouth, but it had the taste of soil. They continued to climb but by seven o'clock in the evening found that they were only halfway to the peak. The sun had gone behind the mountain and only a short span of daylight remained. They sat down to discuss what they should do. All agreed that it would get much colder and if they stayed on the mountain the three of them might well die of exposure. On the other hand, if they simply slid back down, the whole climb was for nothing. To get to the top or find the tail with the batteries was the only chance of survival for all twenty-seven. They made up their minds to remain on the mountain for the night and look for an outcrop of rocks which would provide some shelter.

A little farther up they found a small hillock where the snow had been blown away to reveal the rocks underneath. They piled up loose stones to form a windbreak and, as dark was almost upon them, lay down to sleep. With the dark, as always, came the cold, and for all the protection their light clothes afforded them against the subzero wind, they might as well have been naked. There was no question of sleep. They were compelled to hit one another with their fists and feet to keep their circulation going, begging one another to be hit in the face until their mouths were frozen and no words would come from them. Not one of the three thought he would survive the night. When the sun eventually rose in the east, each one was amazed to see it, and as it climbed in the sky it brought a little warmth back to their chilled bodies. Their clothes were soaked through, so they stood and took off their trousers, shirts, and socks and wrung them out. Then the sun went behind a cloud so they dressed again in their wet clothes and set off up the mountain.

Every now and then they stopped to rest and glance back toward the wreck of the Fairchild. By now it was a tiny dot in the snow, indistinguishable from any of the thousand outcrops of rock unless one knew exactly where to look. The red S which some of the boys had painted on the roof was invisible, and it was clear to the three why they had not been rescued: the plane simply could not be seen from the air. Nor was this all that depressed them. The higher they climbed, the more snow-covered mountains came into view. There was nothing to suggest that they were at the edge of the Andes, but they could only sce to the north and the east. The mountain they were climbing still blocked their view to the south and west, and they seemed little nearer its summit. Every time they thought they had reached it, they would find that they were only at the top of a ridge; the mountain itself still towered above them.

At last, at the top of one of these ridges, their efforts were rewarded. They noticed that the rocks of an exposed outcrop had been broken, and then they saw scattered all around them the twisted pieces of metal that had once been part of the wingspan. A little farther up the mountain, where the ground fell into a small plateau, they saw a seat facedown in the snow. With some difficulty they pulled it upright and found, still strapped to it, the body of one of their friends. Hs face was black, and it occurred to them that he might have been burned from the fuel escaping from the engine of the plane.

With great care Zerbino took from the body a wallet and identity card and, from around the neck, a chain and holy medals. He did the same when they came across the bodies of the three other Old Christians and the two members of the crew who had fallen out the back of the plane.

The three now made a count of those who were there and those who were below, and the tally came to forty-four. One body was missing. Then they remembered the floundering figure of Valeta, who had disappeared in the snow beneath them on that first afternoon. The count was now correct: six bodies at the top of the mountain, eleven down below, Valeta, twenty-four alive in the Fairchild, and the three of them there. All were accounted for.

They were still not at the summit, but there was no sign of the tail section or any other wreckage above them. They started back down the mountain, again following the track made by the fuselage, and on another shelf on the steep decline they found one of the plane's engines. The view from where they stood was majestic, and the bright sunlight reflecting off the snow made them squint as they observed the daunting panorama around them. They all had sunglasses, but Zerbino's were broken at the bridge, and as he climbed the mountain they had slipped forward so that he found it easier to peer over them. He did the same as they started to slide down again, using cushions they had taken from the seats at the top as makeshift sleds. They zigzagged, stopping at each piece of metal or debris to see if they could find anything useful. They discovered part of the plane's heating system, the lavatory, and fragments of the tail, but not the tail itself. Coming to a point where the track of the fuselage followed too steep a course, they crossed to the side of the mountain. By this time, Zerbino was so blinded by the snow that he could hardly see. He had to grope his way along, guided at times by the others. "I think," said Maspons, as they approached the plane once again, "that we shouldn't tell the others how hopeless it seems."

"No," said Turcatti. "There's no point in depressing them." Then he said, "By the way, what's happened to your shoe?"

Maspons looked down at his foot and saw that his shoe had come off while he was walking. His feet had become so numb with cold that he had not noticed.

The twenty-four other survivors were delighted to see the three return, but they were bitterly disappointed that they had not found the tail and appalled at their physical condition. All three hobbled on frozen feet and looked dreadful after their night out on the mountainside, and Zerbino was practically blind. They were immediately taken into the fuselage on cushions and brought large pieces of meat, which they gobbled down. Next Canessa treated their eyes, all of which were watering, with some drops called Colirio which he had found in a suitcase and thought might do them good. The drops stung but reassured them that something was being done for their condition. Then Zerbino bandaged his eyes with a rugby shirt, keeping it on for the next two days. When he removed his bandage he could still see only light and shadow, and he kept the rugby shirt as a kind of veil, shielding his eyes from the sun. He ate under the veil, and his blindness made him intolerably aggressive and irritable.

Their feet had also suffered. They were red and swollen with the cold, and their friends massaged them gently. It escaped no one's notice, however, that this expedition of a single day had almost killed three of the strongest among them, and morale once again declined.

5

On one of the days which followed, the sun disappeared behind clouds, rendering the water-making devices useless, so the boys had to return to the old method of putting the snow in bottles and shaking them. Then it occurred to Roy Harley and Carlitos Páez to make a fire with some empty Coca-Cola crates they had found in the luggage compartment of the plane. They held the aluminum sheets over the fire, and water was soon dripping into the bottles. In a short time they had enough.

The embers of the fire were still hot; it seemed sensible to try cooking a piece of meat on the hot foil. They did not leave it on for long, but the slight browning of the flesh gave it an immeasurably better flavor—softer than beef but with much the same taste.

The aroma soon brought other boys around the fire, and Coche Inciarte, who had continued to feel the greatest repugnance for the raw flesh, found it quite palatable when cooked. Roy Harley, Numa Turcatti, and Eduardo Strauch also found it easier to overcome their revulsion when the meat was roasted and they could eat it as though it were beef.

Canessa and the Strauch cousins were against the idea of cooking the meat, and since they had gained some authority over the group, their views could not be ignored. "Don't you realize," said Canessa, knowledgeable and assertive as ever, "that proteins begin to die off at temperatures above forty degrees centigrade? If you want to get the most benefit from the meat, you must eat it raw."

"And when you cook it," said Fernández, looking down on the small steaks spitting on the aluminum foil, "the meat shrinks in size. A lot of its food value goes up in smoke or just melts away."

These arguments did not convince Harley or Inciarte,

who could hardly derive nutrition from raw meat if they could not bring themselves to eat it, but in any case the limit to cooking was set by the extreme shortage of fuel—there were only three crates—and the high winds which so often made it impossible to light a fire out in the snow.

In the next few days, after Eduardo Strauch became very weak and emaciated, he finally overcame his revulsion to raw meat—forced to by his two cousins. Harley, Inciarte, and Turcatti never did, yet they were committed to survival and managed to consume enough to keep alive. The only ones who still had not eaten human flesh were the two eldest among them, Liliana and Javier Methol, and as the days passed and the twenty-five young men grew stronger on their new diet, the married couple, living on what remained of the wine, chocolate, and jam, grew thinner and more feeble.

The boys watched their growing debility with alarm. Marcelo begged them over and over again to overcome their reluctance and eat the meat. He used every argument, above all those words of Pedro Algorta. "Think of it as Communion. Think of it as the body and blood of Christ, because this is food that God has given us because He wants us to live."

Liliana listened to what he said, but time and again she gently shook her head. "There's nothing wrong with you doing it, Marcelo, but I can't, I just can't." For a time Javier followed her example. He still suffered from the altitude and was cared for by Liliana almost as though he were her child. The days passed slowly and there were moments when they found themselves alone; then they would talk together of their home in Montevideo, wondering what their children were doing at that hour, anxious that little Marie Noel, who was three, might be crying for her mother, or that their ten-year-old daughter, María Laura, might be skipping her homework.

Javier tried to reassure his wife that her parents would have moved into their house and would be looking after the children. They talked about Liliana's mother and father, and Liliana asked whether it would be possible when they returned to have her parents come and live in their house in Carrasco. She looked a little nervously at her husband when she suggested it, knowing that not every husband likes the idea of his parents-in-law living under the same roof, but Javier simply smiled and said, "Of course. Why didn't we think of that before?"

They discussed how they might build an annex onto the house so that Liliana's parents could be more or less independent. Liliana worried that they might not be able to afford it or that an extra wing might spoil the garden, but on every point Javier reassured her. Their conversation, however, weakened his resolution not to eat human meat, and so when Marcelo next offered him a piece of flesh, Javier took it and thrust it down his throat.

There remained only Liliana. Weak though she was, with life ebbing from her body, her mood remained serene. She wrote a short note to her children, saying how dear they were to her. She remained close to her husband, helping him because he was weaker, sometimes even a little irritable with him because the altitude sickness made his movements clumsy and slow, but with death so near, their partnership did not falter. Their life was one, on the mountain as it had been in Montevideo, and in these desperate conditions the bond between them held fast. Even sorrow was a part of the bond, and when they talked together of the four children they might never see again, tears not only of sadness but of joy fell down their cheeks, for what they missed now showed them what they had had.

One evening just before the sun had set, and when the twenty-seven survivors were preparing to take shelter from the cold in the fuselage of the plane, Liliana turned to Javier and told him that when they returned she would like to have another baby. She felt that if she was alive it was because God wanted her to do so.

Javier was delighted. He loved his children and had always wanted to have more, yet when he looked at his wife he could see through the tears in his eyes the poignance of her suggestion. After more then ten days without food the reserves had been drawn from her body. The bones protruded from her cheeks and her eyes were sunk into their sockets; only her smile was the same as before. He said to her, "Liliana, we must face up to it. None of this will happen if we don't survive."

She nodded. "I know."

"God wants us to survive."

"Yes. He wants us to survive."

"And there's only one way."

"Yes. There's only one way."

Slowly, because of their weakness, Javier and Liliana returned to the group of boys as they lined up to climb into the Fairchild.

"I've changed my mind," Liliana said to Marcelo. "I will eat the meat."

Marcelo went to the roof of the plane and brought down a small portion of human flesh which had been drying in the sun. Liliana took a piece and forced it down into her stomach.

Four

1

The news that the Chileans had abandoned the search for their sons after only eight days, on two of which all planes had been grounded because of bad weather, appalled those parents who were still convinced that their sons were alive. They felt bitterly disappointed in the Chileans for giving up so easily and angry with their own government for doing so little. In Chile, Páez Viló announced that his search would go on; in Carrasco, Madelón Rodríguez contacted Gerard Croiset.

Gerard Croiset was born in 1910 of Dutch-Jewish parents and as a young man had been practically illiterate. In 1945 he was discovered by a man called Willem Tenhaeff, who had began to do systematic research into the phenomenon of clairvoyance. In 1953, at the University of Utrecht, Tenhaeff was made Professor of Parapsychology—unprecedented as an accredited subject—and most astounding among the forty or so clairvoyants who worked with him was Croiset.

Croiset's most notable talent was for finding missing people, and for this reason he had often been consulted by the police in Holland and the United States. His method was to handle an object which had belonged to the missing person, or talk to someone involved, and then describe the picture or series of pictures that formed in his mind. If a case resembled some experience of his own he found that his psychic sense was keener; thus if a missing child had been drowned in a canal he was more likely to see it because he himself had almost drowned when he was young. He never accepted money for the use of his powers, and he responded less successfully to those who called on him simply to recover their property.

Every case on which he worked was documented by Professor Tenhaeff, and after almost twenty years and several hundred experiments he had built up an impressive record of success.

Madelón went to the Netherlands embassy in Montevideo and, using one of their staff as an interpreter, telephoned the Parapsychological Institute in Utrecht. She was told that Gerard Croiset was in a hospital recuperating after an operation. She pleaded nevertheless to be put in touch with him and eventually was connected to his son, Gerard Croiset, Jr., in Enschede, who was thirty-four years old and was thought to have inherited his father's powers. Through the interpreter young Croiset asked for a map of the Andes to be sent to him.

Madelón immediately dispatched an aeronautical chart of the area, with a rudimentary diagram marking the air corridors in Chile and Argentina. Arrows had been drawn on the chart, showing the route of the Fairchild, with a question mark on the point of the Planchon Pass.

When she next telephoned young Croiset he told her that he had been in contact with the plane. One of its engines had broken down, and it had lost height. The pilot had not been flying the plane, but the copilot had crossed the Andes before and remembered a valley where he thought he could make an emergency landing. He had therefore turned to the left (the south) or possibly to the right (the north) and had crashed by a lake forty-one miles from Planchon. The plane lay "like a worm"; its nose was crushed. He could no longer see the pilots but he could see life. There were survivors.

Madelón knew that a Japanese clairvoyant living in Córdoba in Argentina had said that the plane had flown south. This seemed to her to confirm the choice of south, not north, from Planchon. She went at once to the house of Rafael Ponce de León.

Rafael, too, had been shocked when the search was called off, and he had made up his mind that while any of the parents still looked for their sons he would see to it that they had at their command a whole network of communication. For this purpose he kept in contact with a number of other radio hams in Chile. Through one of them he reached Páez Vilaró, and Madelón told him of her conversation with Gerard Croiset, Jr.

The news that the Dutch clairvoyant had made contact with the plane spread rapidly among the other parents. Although many of them treated it with skepticism, especially the fathers, they nevertheless appointed a delegation of three to go to the commander in chief of the Uruguayan Air Force. The delegation made a formal request for a Uruguayan aircraft to be sent to Chile to search for the Fairchild in the mountains around Talca, a town lying about 150 miles south of Santiago. This request was refused.

The news of the young Croiset's vision did much to raise Páez Vilaró's morale. He had always found magic more impressive than science. He had also flown over the area where the SAR thought the plane had come down, the Tinguiririca and Palomo volcanoes, and knew that there was nothing he could do to search among the mountains at such an altitude; but Croiset had put the site of the accident in the precordillera, where the mountains were much lower. A labor of Hercules had been reduced to mortal proportions.

Páez Vilaró immediately set out for the south, and next day, Sunday, October 22, he was flying over the mountains around Talca in a plane obtained at the Aero Club of San Fernando.

In the days which followed he lived in a frenzy of activity. He made a list of all those who owned private planes in Chile and asked the pilots for advice, which invariably became an offer of their services. Páez Vilaró could have had thirty planes at his disposal, and he only hesitated to use them because of the acute fuel shortage in Chile. He knew that the man who took him up in his plane for an hour would be sacrificing the use of his car for a month; yet many people, though personally convinced the boys were dead, did this without asking for payment.

Next to be organized were the radio hams whom Rafael had recruited over the air from Carrasco. Many of these men offered Páez Vilaró not only their radios but their clothes and cars. Wherever he went in the

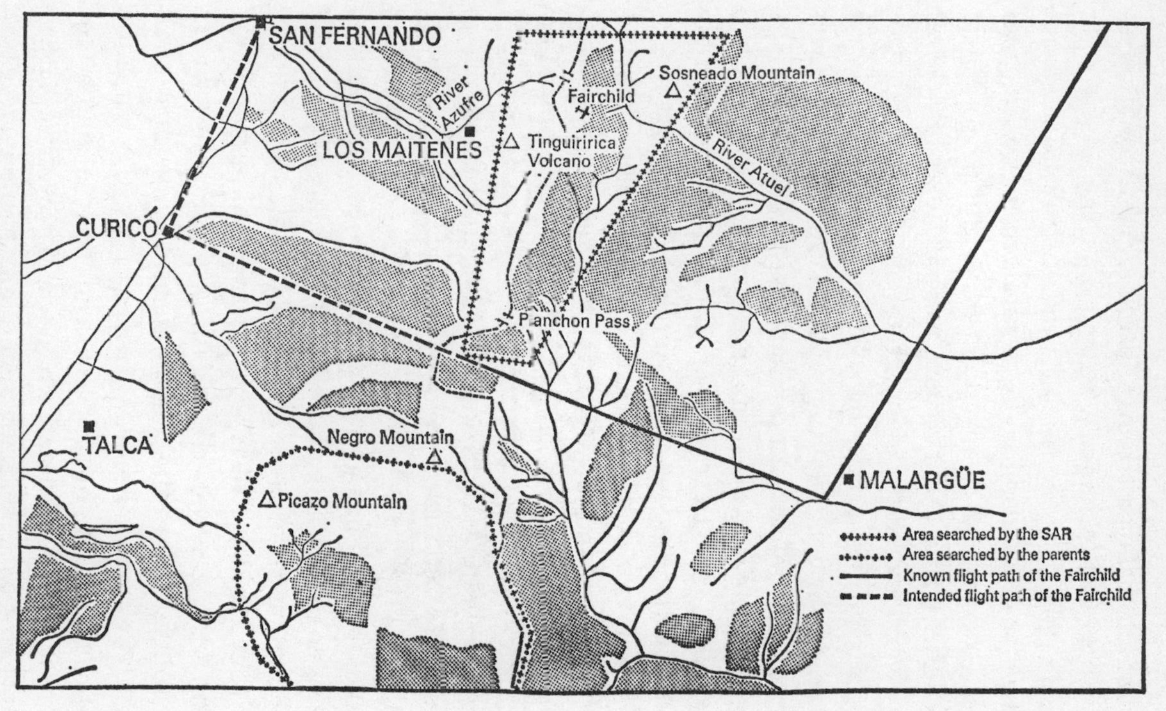

The area of the searches

mountains he would be followed by a Citroen Deux Chevaux with antennas waving like the horns of a grasshopper. The car could put him in touch with Rafael in Montevideo at a moment's notice and, through Rafael, with anyone in the world.

Páez Vilaró did not remain in Talca itself but set off on several expeditions into the Andes. Madelón Rodríguez and the mother of Diego Storm had arrived in Talca, which freed him to pursue his own plans. He was not content to have just the rich Chileans with their private airplanes help him find the boys; he wanted word to get around to the poorest peasant in the most remote valley of the Andes that a search was on for the survivors. In every village he came to, he asked if anyone had seen a plane fall from the sky, and he listened to many fascinating but irrelevant stories. He entertained those he questioned with a drink or a cup of coffee. At one time he had four rooms in four different hotels, in case the search took him in any one of those four directions. He still had no money, but the innkeepers and restaurateurs either wanted no money or were paid with a drawing on a plate, napkin, or tablecloth.

His reputation preceded him. Now, when he entered a village, a small crowd would gather and people would shout, "Here comes the lunatic who's looking for his son!" Páez Vilaró did not mind; he saw his mission as something magical and fantastic, an army deployed in search of a plane under the directions of a Dutch seer. The villagers considered him a wizard because he carried with him a Polaroid camera and would give gifts of their images to men who had never seen a photograph before.

By plane and on foot he scoured the area forty-one miles from the air lane over Planchon, yet nothing was found. He asked Rafael over the radio to ring Croiset once again and ask for more details. This was done, and night after night at 2 A.M. Croiset, in his pajamas, would answer the telephone and summon up images of the Andes.

He was able to provide them with more details, but much of what he saw concerned the flight of the plane rather than its present situation. In stages, he described a fat man—probably the pilot—with food poisoning, and how he went from the cabin and left the controls in the hands of the copilot. He was wearing a blazer and fiddling with his spectacles. Then the plane's engine went dead and the copilot flew the plane toward a beach, perhaps on the sea, perhaps on a lake, which he remembered from previous crossings of the Andes. He had found a lake—or, rather, a group of three lakes—and attempted to land, but the plane had crashed into the base of the mountain and was concealed by an overhanging shelf of rock. Near it there was a mountain "without a top," and he saw danger—perhaps a road sign saying danger. He could no longer see life in the plane, but perhaps this only meant that the boys had left it and had taken shelter nearby.

With each new detail Páez Vilaró and his Chilean friends set out again to search in the mountains, and by this time faith in Croiset's vision had spread to others. Madelón had gone to Santiago and persuaded the SAR to send planes over the mountains around Talca. The commander of the army in Talca sent out a patrol toward the Cerro Picazo (which satisfied the description of a mountain without a top), and for five days these Chilean soldiers searched in the bitter cold for the wreck of the Fairchild. A group of Silesian priests also set out into the mountains and searched for three days among inaccessible mountains as a "prayer of hope."

Nothing was found, and since low flying over the mountains was exceptionally dangerous, the SAR again called off the search. Only helicopters could fly low enough to see a plane half hidden by a mountain or Uruguayan boys sheltering under pine trees, and at a time when soap and cigarettes were unobtainable in Chile, helicopters were practically impossible to come by.

But to Madelón, this was only a minor obstacle, and she decided to apply to President Salvador Allende himself for the use of his personal helicopter. Before she did so, however, a friend told her of an acquaintance who hired out small "choppers" for spraying crops and raising high-voltage cables onto pylons, and in ten minutes it was arranged that when these helicopters were available, Madelón should hire them at the charitable rate of ten dollars an hour.

Meanwhile Páez Vilaró and Rafael agreed, on the afternoon of October 28, that as much as possible had been done without helicopters to follow up the vision of Gerard Croiset.

2

On Sunday, October 29, the anniversary of the death of Marcelo Pérez's father, his mother, Estela, invited the parents of the boys who had gone in the Fairchild to come to her house that afternoon for a meeting. More than parents came; there were also brothers, sisters, and *novias* of many of the Old Christians.

The table in Estela's spacious living room was covered with maps of the Andes, with circles and lines drawn around Talca to show what areas had already been covered by Páez Vilaró. On a sideboard, someone had put a pile of mushrooms which were known to grow in the cordillera and were perhaps the food on which their sons were now living.

The atmosphere in the room was black. There was none of the optimism that had been shown the week before, when Croiset had first given them the fruit of his clairvoyance. By their fidgeting mannerisms and blurted speech, many of the women, especially the girls, showed themselves to be on the brink of hysteria. Others simply sat in the stunned silence of their despair.

Estela started the meeting. "I have asked you here," she said, "because I feel that we must do something. We can't just sit here in Montevideo, waiting—"

"Páez Vilaró is searching," said a voice from among the group of parents.

"Yes," said Estela. "One man in the whole cordillera."

"I really don't think that there's much more to be done," said one of the fathers.

At this, one of the girls said contemptuously, "It looks as if Páez Vilaró is the only father for all those boys. No one else is with him . . ." She paused, then spat out, "Or must we women go to Chile?"

The room exploded with different voices and different opinions. When finally there was quiet, Jorge Zerbino, a lawyer and businessman, turned to Luis Surraco, a doctor, and said, "I'm going to Chile, Luis. Do you want to come with me?"

Dr. Surraco, the father of Roberto Canessa's *novia*, was adept at reading maps, and they listened to him describe where the two might search. The views of the majority remained that they should follow the instructions of Croiset. The clairvoyant had sent some new scraps of information, and though Zerbino and Surraco both had grave misgivings about Croiset, they knew quite well that their expedition was not so much to find the boys as to keep up the spirits of the women at home. They agreed, therefore, to search around Talca.

When the meeting broke up, Rafael Ponce de León got in touch with Páez Vilaró over the radio. "Zerbino and Surraco are coming to Chile," he said. "They're coming to help you."

The voice which answered did not respond to his optimistic tone. "They shouldn't bother," said Páez Vilaró despondently. "There isn't much point."

Rafael was astonished. How, after all this, could Carlos not want them to come?

"Are you alone?" the painter asked him.

"Yes," replied Rafael.

"Don't tell the others," said Páez Vilaró, his voice slow and heavy, "but it's hopeless. I've lost all faith in finding them, my boys. I still search for them with the Cross in one hand and the signs of the zodiac in the other, but I don't believe that they're alive any more."

There was a pause; then Rafael said, "Come back, Carlos. Everyone would understand if you come back."

"No," said Páez Vilaró. "Madelón still believes they

are alive. I can't leave her to despair." Rafael could hear him sobbing into the microphone.

The next day Rafael told Dr. Zerbino and Dr. Surraco something of this exchange, but they refused to be discouraged. They spoke to Páez Vilaró over the radio, telling him that they had already booked seats on a plane to come to Chile, and Páez Vilaró did not rebuff them. "Come on out," he said with his usual warmth. "I'll be waiting for you."

The crowd around the radio transmitter became alive with excitement. Help, advice, money, and maps were pressed upon Zerbino and Surraco. The fathers of Daniel Shaw and Roy Harley formed a fund to finance the search, and contributions came in even from those men like Seler Parrado who were sure that their sons were dead. Parrado had been among the most desolate of those affected by the accident. His despair was total. Not only had he lost the wife upon whom he had depended so much but two of his children as well. He had toiled all his life to build up a business, not for his own satisfaction but for the sake of his family. Now they were gone. His one surviving daughter had moved into her father's house to look after him, but the life had gone from Seler Parrado's eyes. He could see no reason to go on; there was nothing left to live for. With a dead heart he had sold Nando's Suzuki motorcycle to a boy who had once been his friend. Still, he felt he wanted to contribute to the fund.

That night it was announced on the radio that because of unprecedented falls of snow in the Andes, the SAR would not resume their search for the wreck of the Fairchild in January, as previously announced, but would wait until February. The news did not daunt Zerbino and Surraco. Their suitcases were packed. They would leave the next day.

Dr. Zerbino and Dr. Surraco, joined by Guillermo Risso, a friend of Gaston Costemalle, flew to Santiago on November 1. There they met Madelón and Storm's mother, who had left Talca and were on their way to Córdoba in Argentina to try and bring the Japanese clairvoyant to Santiago. The two women told them what had happened in Talca, and in the afternoon Zerbino and Surraco and Risso continued their journey in a hired car. The political situation in Chile had deteriorated still further, and the enemies of Allende's government had strewn the streets leading out of Santiago with hooks and nails in an attempt to bring all traffic to a standstill. The result was that the car in which the three Uruguayans were traveling suffered a series of punctures which greatly hampered its progress.

Páez Vilaró was waiting for them in jail. That morning he had flown over a power station, and the local police, nervous no doubt because of the political situation, had decided that he and the young Uruguayan who was with him, a friend of the boys, were spies. They were treated politely; Páez Vilaró was able to make contact with Ponce de León, to whom he tried to communicate his predicament without actually describing it because he had no wish to trouble the women who he knew would be crowding around Rafael's radio transmitter. "I'm sitting here," he said, "with a lot of little bars in front of me . . . there's a beautiful view just like that from the Punta Garreta"—the chief prison in Montevideo.

By late afternoon, however, the police had discovered that Páez Vilaró was not an agent of a foreign power but "the lunatic who's looking for his boy," and the two Uruguayans were released just in time to welcome Zerbino, Surraco, and Risso. Páez Vilaró at once took them on a tour of inspection and surprised them with the extent and efficiency of his organization. That evening they were able to speak to Croiset in the Netherlands through the link of Ponce de León in Montevideo and were given another clue for the search: In the lake where the plane had crashed there was an island.

Páez Vilaró remembered that on a former flight in the hired helicopter he had seen just such a lake about sixty miles from Talca. The next day he flew to it again and searched in the area to the right of the two mountains which had no peaks, the Cerro Azul and the Cerro Picazo. They flew down canyons and into boxes between mountains, but the result was always negative.

The next day the Uruguayans wanted to search the same area, but the helicopter had to return to Santiago. "Anyway," the pilot of the helicopter told them, "if the plane had crashed in country like that, it would have been buried in the snow." They were not discouraged. Under Dr. Surraco's direction, they built a cardboard model of the area and then inquired about professional mountaineers who might be engaged to assist them.

That evening, they spoke as usual with Carrasco and were told that there had been a mistake in the translation of one of Croiset's messages. The plane would be found to the left of a mountain without a peak, not to the right, as they had previously been told.

They therefore set off the next day, the third of November, with two guides toward the small town of Vilches, thirty-seven miles from Talca. There they divided into two groups and went into the mountains.

The fathers were not as fit as their sons, and it was not easy for them to climb in the cold air with packs on their backs. Both groups climbed the Cerro del Peine, but when they reached the summit nothing could be seen. In a thick cloud, they began the descent to Vilches, clambering over boulders and steadying themselves with thick staves as their knees wobbled on the steep decline.

The Laguna del Alto lay on their way back, and they explored the rocks around it for signs of a plane. There were none.

They were back in Vilches by November 7, and on the eighth, the helicopter was free again and returned from Santiago. In the morning it flew over the Despalmado mountain; in the afternoon it checked the area of the Quebrada del Toro, where a peasant was reported to have heard the sound of a plane crashing to the ground. The Uruguayans waited in Vilches for the results of these forays; they were all negative.

On November 9 the party returned to Talca and on the tenth to Santiago. They reported to the SAR on what they had done, and the authorities repeated that the official search would not be resumed until the thaw had set in—"at the end of January, perhaps, or the beginning of February"—and then in the area of the Tinguiririca volcano.

On the same day, in Montevideo, news came that Croiset had made a drawing of the area of the accident and had recorded on tape a more extensive description than he had been able to give over the telephone. The package containing them would arrive via KLM at noon the next day.

Some time before it did so, a group of relatives gathered at Carrasco airport to await this important package. There was a flight to Santiago that afternoon, and they wanted to have Croiset's drawings copied and his tape transcribed before sending the originals on to Páez Vilaró, Zerbino, and Surraco in Santiago on that flight.

With them at the airport were the consul of the Netherlands and the father of an Old Christian, Bobby Jaugust, who was also the representative of KLM in Uruguay. To have brought these two men proved a wise precaution, for the parcel from Croiset had not been shipped separately but was somewhere in two sacks of mail from Europe. Authority was given to open and search them, and eventually the parcel was found. It was opened at once; one team set to work copying his sketches and another to transcribing the tape.

When this had been done, the original tape and drawings were put on the SAS flight to Santiago. Páez Vilaró was then contacted by radio and told of the conclusions they had reached: Everything in the drawings and the message pointed to the Laguna del Alto in the precordillera by Talca.

The three men in Santiago were less convinced. After days searching in the precordillera they were in a better position to assess the value of Croiset's package, and much of what he had said seemed quite irrelevant to the conditions they had encountered.

He said the accident had occurred near a beach, either by the sea or on a lake. Close by there was a shepherd's hut and only a little farther a village with white Mexican-style houses, near which a battle had taken place in 1876. He saw letters and figures on the plane—an N and a Y and the figures 3002. The figure 1036 had also come into his head, perhaps signifying that the plane was 1,036 meters (3,400 feet) above sea level.

The nose of the plane was crushed; it had come down softly like an insect and had lost both its wings. He saw the fuselage separately from the rest of the plane but could not identify its markings, perhaps because it was too dark under the shelf of rock where it had crashed. Nor could he see any life in the plane; no one looked out of the windows.

His sketches were rudimentary. There was also a triangle giving specific distances but no point on which they could take a bearing. In all, it was a mixture of magic and technical data which Surraco, for one, could not stomach. "It's completely unscientific," he said. "We're chasing after nothing at all. If we search anywhere it should be around the Tinguiririca volcano. That's where the facts as we know them tell us the plane should be."

Zerbino agreed with him. He saw no point in returning to Talca, and since they had not the means to search in the higher mountains around the Tinguiririca, he booked seats for himself and Surraco to return to Montevideo the next day.

Páez Vilaró temporized. He certainly had doubts about Croiset, but he could not bring himself to disappoint Madelón and the other women who still believed in him. He therefore told Zerbino and Surraco that he would remain in Chile for a day or two more, and when they returned to Uruguay he went back to Talca. There he made one more trip to the Laguna del Alto but found nothing.

Some time before the boys had left for Chile, Páez Vilaró had made a commitment to go to Brazil in the middle of November. It was now close to the day when he was expected in São Paulo, so he prepared to leave. He had spent more than a month hunting for the plane, and even now he made arrangements for others to continue the search while he was away. He had printed several thousand leaflets offering, on behalf of the parents, a reward of 300,000 escudos to anyone who gave information leading to the finding of the Fairchild. He also prepared the ground for Estela Pérez, who was to take over in Talca, and before he went he gave some money to the schoolchildren of Talca to enable them to form a soccer team called the Old Christians.

On November 16, Páez Vilaró returned to Montevideo.

Five

1

The seventeenth day, October 29, passed quite well for those stranded in the Fairchild. They were still cold, wet, dirty, and hungry, and some were in great pain, but in the last few days a degree of order seemed to have been imposed on the chaos. The different teams for cutting, cooking, melting snow, and cleaning the cabin were working well, and the wounded were sleeping a little more comfortably in the hanging beds. More important still, they had started to single out the fittest among them as potential expeditionaries who would master the Andes and get help. Their mood was optimistic.

They ate at midday; by half past four in the afternoon the sun went behind the mountains to the west, and at once it became bitterly cold. They filed in groups of two into the hulk of the plane in the order they were to sleep—Juan Carlos Menéndez, Pancho Delgado, Roque, the mechanic, and Numa Turcatti entered last, for it was their turn to sleep by the entrance.

Each boy, as he entered, took off his shoes and put them up on part of the hat rack on the right-hand side.

They had decided that day to make this rule to save the cushions and blankets from getting wet. Then the couples crawled up the plane to their assigned places.

Though it was only the middle of the afternoon, some closed their eyes and tried to sleep. Vizintín had slept badly the night before and was determined to make himself as warm and comfortable as possible. He had been allowed to keep his shoes on because he slept exposed to the cold on the hanging bunk. There was a strong wind outside which blew freezing air through every hole and crack in the plane. He had managed to secure a large number of cushions and blankets (the covers of the cushions which they had sewn together), and with these he padded and covered his body, including his head.

Carlitos Páez said the rosary out loud, and some of the boys talked quietly among themselves. Gustavo Nicolich confided to Roy Harley his hope that if he died someone would take back the letter he had written to his *novia*. "And if we all die," he said, "they might find the wreck and the letter and give it to her. I miss her so much I feel so awful because I paid very little attention to her. And to her mother." There was a silence; then he added, "There are so many things one regrets. . . . I hope I have a chance to put them right."

The dim light grew dimmer still; a few drifted into half-sleep, and their breathing took on a more regular pattern which in its turn lulled others to sleep. Canessa remained awake, trying to communicate telepathically with his mother in Montevideo. He held a strong image of her in his mind and repeated over and over again in emphatic whispers which were inaudible to the others,

"Mama, I am alive, I am alive, I am alive. . . ." Eventually he too dozed off.

The plane was now silent, but Diego Storm could not sleep because of a painful sore on his back. He was lying between Javier Methol and Carlitos Páez on the floor, but the longer he lay in such discomfort, the more convinced he became that it would be better on the other side of the plane. He looked across and saw that Roy Harley was still awake, so he asked him if he would change places. Roy agreed and they squeezed out of their positions and crawled by one another.

Roy lay down on the floor with a shirt covering his face, thinking about what Nicolich had said, when he felt a faint vibration and an instant later heard the sound of metal falling to the ground. This sound made him jump up, but as he did so he was smothered in snow. He found himself standing buried up to his waist and when he took the shirt from his eyes what he saw appalled him. The plane was almost entirely filled with snow. The wall at the entrance had been toppled and buried, and the blankets, cushions, and sleeping bodies which had covered the floor were now hidden. Quickly, Roy turned to his right and burrowed for Carlitos, who had been sleeping there. He uncovered his face, then his torso, but still Carlitos could not free himself. There was a creak as the snow settled, and in the bitter cold its surface immediately began to form into brittle ice.

Roy left Carlitos because he saw the hands of others sticking out of the snow. He felt desperate; he alone seemed to be free to help. He uncovered Canessa and then went to the front of the cabin and dug out Fito Strauch, but the minutes were passing and many boys remained buried. Above, from one of the hanging beds, Vizintín had started to burrow in the snow, but Echavarren could not move and Nogueira, though free, seemed paralyzed by shock.

Roy crawled frantically to the entrance and squeezed himself out of the small hole that was left, as if he might shovel out the snow by the way it had come in, but he realized at once that this was hopeless so he crawled back into the cabin. There he saw that Fito Strauch, Canessa, Páez, and Moncho Sabella were free and digging.

Fito Strauch had been talking with Coche Inciarte when the avalanche fell upon them. He realized immediately what had happened and struggled against the grip of the snow, but he could not move any part of his body so much as an inch one way or the other. He relaxed and thought with resignation that he was about to die; even if he could escape he might be the only one to do so, and perhaps it would be better to die than to survive alone, isolated in the Andes. Then he heard voices and Roy Harley took hold of his hand. As Roy burrowed toward his face, Fito told his cousin Eduardo, through a hole between them in the snow, to keep calm, to breathe slowly, and to ask after Marcelo. After that he felt a sharp pain in his toe and realized that Inciarte had bitten it. He too was alive.

Fito was freed. Eduardo climbed out the same hole, and Inciarte, after digging a short tunnel, emerged, followed by Daniel Fernández and Bobby Francois. They all immediately began to burrow with their bare hands in the packed snow, and the first they dug for was

Marcelo. When they found his face, however, they saw that he was already dead.

Fito now worked hard to dig down to the living. He also organized the others, who were in some cases so dazed that they did not know what they were doing. Even when a stitch forced him to rest, he continued to urge on the different teams so that those who were digging one hole would not throw their snow into a shaft dug by others.

Parrado lay in the middle of the plane with Liliana Methol on his left and Daniel Maspons on his right. He heard and saw nothing but suddenly found that he was smothered and paralyzed by heavy, cold snow. He could not breathe, but he had read in *The Reader's Digest* that it was possible to live under the snow, so he attempted to take small breaths. He continued to do this for several minutes, but the weight on his chest became more terrible, he grew dizzy, and he knew that he was about to die. He did not think of God or of his family but remarked to himself, "Okay, I'm dying." Then, just as his lungs were about to explode, the snow was scraped from his face.

Coche Inciarte had seen the avalanche and then heard it, a *woosh* followed by silence. He lay immobilized with three feet of snow over him and Fito's toe in his face. He bit it. It was the only way to find out if Fito was alive or tell him that he was. The toe moved.

The snow settled on top of him and its weight made him urinate. He could not breathe or move. He waited and then felt the toe being dragged away from his face.

He struggled against the snow and finally slipped out by the same tunnel.

Carlitos Páez had been uncovered to the waist by Roy but still couldn't move until Fito, when freed, dug away the snow from around his legs. He immediately began to look for his friends Nicolich and Storm, but the snow froze his hands as he burrowed. He warmed them quickly with his gas cigarette lighter and continued to dig, but when he found Nicolich and gripped his hand it was cold and lifeless and gave no clasp in return.

There was no time for lamentation. Carlitos at once dug the snow away from Zerbino's face and then freed Parrado. Then he turned and burrowed toward Diego Storm, but the snow he scraped away fell onto Parrado, who swore at him. He dug more carefully, but it was all to no avail; when he found him, Diego was dead.

To Canessa, the avalanche came like the magnesium flash of an old camera. He too was buried, imprisoned and suffocating, and like Parrado he was possessed less by panic than by curiosity. Well, he thought to himself, I've got as far as this, and now I'm going to know what it's like to die. At last I'll experience all those abstract notions like God and Purgatory and Heaven or Hell. I've always wondered how the story of my life would end; well, here I am at the last chapter. Yet just as the book was about to be closed, a hand touched him; he clasped it, and Roy Harley sank a shaft to bring air to his lungs.

As soon as he could move Canessa searched for Daniel Maspons. He found his friend lying as if asleep, but he was dead.

* * *

The snow which covered Zerbino left a small cavity which enabled him to breathe for a few minutes. Like Canessa and Parrado, he did not pray to God or repent of his sins, but though his mind was calm his body was not resigned to death. He had thrown up one arm at the moment the avalanche struck, and his struggles opened a fissure in the snow beside it, down which air came to his lungs.

Above him he heard the gruff voice of Carlitos Páez shout down, "Is that you, Gustavo?"

"Yes!" shouted Zerbino.

"Gustavo Nicolich?"

"No. Gustavo Zerbino."

Carlitos moved on.

Later another voice called down to him, "Are you all right?"

Zerbino replied, "Yes, I'm okay. Save someone else." He then waited in his tomb until the others had time to dig him out.

Roque and Menéndez had been killed by the falling wall, but part of that wall saved the lives of the two others who slept next to it. Numa Turcatti and Pancho Delgado were trapped under the curved door, which had been the emergency exit of the plane and had been built into the wall, but they had air enough to breathe under its concave surface. They survived like this for six or seven minutes. They made sounds, however, and Inciarte came with Zerbino to their rescue. The snow there at the back of the plane was very deep and Inciarte asked Arturo Nogueira, who was watching from his hanging bed, to help them dig. Nogueira did not move, nor did he say anything. He stayed trancelike on the hanging bed.

Pedro Algorta, still buried beneath the snow, had only what air he held in his lungs. He felt himself near to death, yet the knowledge that after his death his body would help the others to survive instilled in him a kind of ecstasy. It was as if he were already at the portals of heaven. Then the snow was scraped away from his face.

Javier Methol had been able to reach out of the snow with his hand, but as they tried to free him he only shouted at the boys to dig toward Liliana instead. Javier could feel his wife with his feet and feared she might be suffocating, but he could do nothing to help her. "Liliana," he shouted, "make an effort! Hold on. I'll get to you!" He knew that she might live for a minute or two without air, but the weight of the boys digging all around him was pressing the snow down upon her. Moreover, their instincts were to help first their own friends, then those whose hands they could see stretching out from beneath the snow. Inevitably, they left until last those like Javier, who could breathe, and those like Liliana, who were completely lost from view.

Javier continued to shout to his wife, begging her to hold on, to have faith, to breathe slowly. Finally, he was freed by Zerbino, and together they dug for Liliana. When they found her she was dead. Javier slumped down onto the snow, weeping, overwhelmed by grief. His only consolation came from his conviction that she who had given him such love and solace on earth must now be watching over him from heaven.

Javier was not alone in sorrow, for when the living huddled together in the few feet of space that was left to them between the roof of the plane and the icy floor of snow, they found that some of their dearest friends lay buried beneath them. Marcelo Pérez was dead; so too were Carlos Roque and Juan Carlos Menéndez, crushed beneath the wall; Enrique Platero, whose stomach wound had healed at last; Gustavo Nicolich, whose courage after the broadcast had saved them from despair; Daniel Maspons, Canessa's closest friend; and Diego Storm, one of the "gang." Eight had died under the snow.

The conditions facing the nineteen who survived were not so terrible that they did not all feel bitter sorrow at the death of their friends. Some wished that they too had been smothered by the snow rather than continue a life of such physical and spiritual suffering without their companions. This wish was almost met, for a second avalanche hit the plane an hour or so after the first, but because the entrance was already blocked, most of this second fall passed over the plane. In doing so, however, it sealed off the gap through which Roy Harley had crawled out and then in again. The Fairchild was completely buried.

As night set in, the survivors were wet, cramped, and bitterly cold with no cushions, shoes, or blankets to protect them. There was barely room to sit or stand; they could only lie in a tangle, punching each other's bodies to keep the blood flowing in their veins, yet not knowing to whom the arms and legs belonged. To make more space, some of the snow in the center of the cabin was shoveled to either end; with the Strauch cousins and Parrado, Roy scooped out a hole with room for four to sit and one to stand. The one whose turn it was to stand would jump on the feet of the others to try and keep them from freezing.

The night was endless. Only Carlitos was able to sleep, and then only for brief periods. The others remained awake, wriggling their fingers and toes and rubbing their faces and hands together to keep warm. After several hours another danger presented itself: The little air that was left in the plane became stale and stuffy. Some of the boys began to feel faint from the lack of oxygen. Roy went to the entrance and tried to dig an air shaft, but his arm could not reach up to the surface, and in any case the snow there had frozen into ice too hard to be penetrated by bare hands. Parrado then took one of the steel poles that had been used to make the hammocks and poked it through the roof of the cabin. He worked by the light of five cigarette lighters as the boys around him watched anxiously, for they had no idea if the snow which covered them was one foot deep or twelve. But after poking the bar through and working it up, Parrado soon felt it slide unimpeded into the fresh air, and when he drew it back into the cabin it left a hole through which he could see the frail light of the moon and stars.

Through this hole they watched for the coming of morning, and eventually the damp blackness inside the plane gave way to a pale, lugubrious light as the sun rose in the east and its rays filtered down through the snow. As soon as they could see what they were doing, they considered how to get out of their tomb. There was too much snow above them to get through at the entrance, but it seemed to lie more thinly over the pilots' cabin; light could be seen filtering through the window. Canessa, Sabella, Inciarte, Fito Strauch, Harley, and Parrado began to tunnel through the pilots' cabin. It was full of frozen snow which they had to remove with their bare hands, and the six worked in turns. Then Zerbino, who wore thick clothing and could stand the cold better than some of the others, squeezed past the dead pilots' bodies and reached the window which, because of the tilt of the plane, looked up toward the sky. He tried to open it but the snow piled on top was too heavy, so he came back down. Canessa tried, but he too failed. Roy went next and finally pushed out the glass and broke through the snow into daylight.

He pushed his head above the surface. It was around eight in the morning but darker than usual because the sky was overcast. Clouds of snow swirled around him. He was warmly dressed in a woolen cap and a waterproof jacket, but the strong wind blew snow into his eyes and stung the skin of his face and hands.

He lowered himself down into the pilots' cabin and shouted down to the others, "It's no good! There's a blizzard out there."

"Try and uncover the windows," someone called.

Roy lifted himself up again and this time climbed out of the plane, but the fuselage behind him was completely covered. It was impossible to see where the windows might be, and he was afraid that if he moved he might slip off the roof and be lost in the snow. He climbed back through the window and rejoined his companions.

The blizzard continued throughout that day, the flakes floating down the tunnel past the corpses of the pilots, still crushed in their seats. The thin layer that collected was scooped up by some of the boys to quench their thirst; others broke off harder lumps of older snow.

It was October 30 and Numa Turcatti's twenty-fifth birthday. The boys gave him an extra cigarette and made a birthday cake out of the snow. Numa was neither an Old Christian nor a rugby player—he had been educated by the Jesuits and preferred soccer—but there was great strength in his stocky figure and calm manner. Many would have liked to give him a better time on his birthday, but instead it was he who improved their spirits. "We have survived the worst," he said. "From now on, things can only get better."

They did nothing that day but suck on snow and wait for the storm to abate. They talked a great deal about the avalanche. Some, like Inciarte, thought that the best of them had died because God loved them most, but others could make no sense of it. Parrado expressed his determination to leave. "As soon as the snow stops," he said, "I'm going. If we wait here any longer, we'll all get killed by another avalanche."

"I don't think so," replied Fito judiciously. "The plane's covered now. The second avalanche went over the top. So we're safe here for the time being. If we start out now, the chances are we'll be hit by an avalanche as we walk through the snow."

They listened to Fito with respect because he had remained calm just after the avalanche, and now he showed none of the hysteria evident in some of the others.

"There's no reason why we shouldn't wait until the weather gets better," he went on.

"But how long?" asked Vizintín, another who wanted to leave at once.

"I remember in Santiago," said Algorta, "a taxi driver told me that the snow stops and summer starts on the fifteenth of November."

"The fifteenth of November," said Fito. "That's just over two weeks. It's worth waiting that long if it adds to your chances of getting through."

No one could argue against this.

"And about that time," he said, "there should be a full moon. It would mean that you could walk by night when the snow's hard and sleep during the day when it's warmer."

They ate nothing that day, and that night, as they huddled together to try and sleep, they all followed Carlitos in the rosary. The next day, October 31, was his nineteenth birthday. The present he would most have wanted, after a cream cake or a raspberry milkshake, was a break in the weather, but when he climbed up the tunnel to the open window the next morning, he saw that it was snowing just as heavily. He returned and predicted to the others, "We'll get three days of bad weather and then three of sunshine."

The bitter cold combined with their wet clothes to deplete their strength. They had eaten nothing for two days and now felt enormously hungry. The bodies of those who had been killed in the crash remained buried in the snow outside the plane, so the cousins uncovered one of those who had been smothered in the avalanche and cut meat off the body right before everyone's eyes. The meat before had either been cooked or at least dried in the sun; now there was no alternative but to eat it wet and raw as it came off the bone, and since they were so hungry, many ate larger pieces, which they had to chew and taste. It was dreadful for all of them; indeed, for some it was impossible to eat gobbets of flesh cut from the body of a friend who two days before had been living beside them. Roberto Canessa and Fito Strauch argued with them; Fito even forced Eduardo to eat the meat. "You must eat it. Otherwise you will die, and we need you alive." But no arguments or exhortations could overcome the physical revulsion in Eduardo Strauch, Inciarte, and Turcatti, and as a result their physical condition deteriorated.

The first of November was All Saints' Day and Pancho Delgado's birthday. As Carlitos had predicted it had stopped snowing, and six of the boys climbed out onto the roof to warm themselves in the sun. Canessa and Zerbino dug the snow off the windows to let more light into the plane, and Fito and Eduardo Strauch and Daniel Fernández melted snow for drinking water, while Carlitos smoked a cigarette and thought about his family, for it was also his father's and sister's birthday. He felt certain now that he would see them again. If God had saved him in both the accident and the avalanche, it could only be to reunite him with his family. The nearness of God in the still landscape set a seal on his conviction.

When the sun went behind a cloud it became cold again, and the six climbed back into the Fairchild. All they could do now was wait.

2

In the days which followed, the weather remained clear. There were no heavy falls of snow, and the stronger and more energetic among the nineteen survivors were able to dig a second tunnel out through the back of the plane. Using shovels made from pieces of metal or plastic broken off the body of the plane, they hacked at the hard snow, recovering objects which had been lost in the avalanche. Páez, for example, found his rugby boots.

Once a tunnel had been made, they were able to set about removing from the cabin both the snow and the bodies buried beneath it. The snow was like rock and their tools were inadequate. The corpses, frozen into the last gestures of self-defense, some with their arms raised to protect their faces like the victims of Vesuvius at Pompeii, were difficult to move. Some of the boys could not bring themselves to touch the dead, especially the bodies of their close friends, so they would tie one of the long nylon luggage straps around the shoulders of the corpse and drag it out.

Those buried inside near the entrance were left there, encased in the wall of ice which protected the living from a further avalanche. They provided a reserve supply of food, in case a second avalanche or heavy blizzard should cover and conceal the bodies they had just taken out, for those who had died in the crash were now completely lost under the snow. For the same reason, when the survivors came in at night, they would leave a limb or a portion of a torso on the "porch," in case the weather the next day made it impossible for them to go out.

It took eight days for the plane to be made more or less habitable, but a wall of snow remained at either end and the space they had to live in was more restricted than before—even allowing for the fewer numbers. Many looked back with mild regret to the halcyon days before the avalanche: "We thought we were badly off then, but what luxury and comfort compared to this!" There was only one advantage to ensue from the avalanche: the extra clothes which could be taken from the dead bodies. Feeling that God would help them if they helped themselves, the survivors not only set about the tasks which would make their immediate life more bearable but planned and prepared for their ultimate escape.

Before the avalanche it had been decided that a party of the fittest among them should set off for Chile. At first there had been a division of opinion between those who thought a larger group would stand a better chance and those who felt that it would be wise to concentrate their resources on a group of only three or four. As it became clear during the weeks following the crash, and especially during the stormy days after the avalanche, that the conditions encountered on any expedition would be severe, the reasoning of the second group prevailed. Four or five would be chosen as expeditionaries. They would be given larger rations of meat and the best places to sleep and be excused from the daily labor of cutting meat and clearing snow, so that when summer finally settled in and the snow began to melt toward the end of November they would be strong, healthy, and fit for their walk to Chile.

The first factor to be considered in choosing these

expeditionaries was their physical condition. Some who had been unharmed in the accident had suffered since. Zerbino's eyes had not fully recovered from his climb up the mountain. Inciarte had painful boils on his leg. Sabella and Fernández were well enough but, not being players, they were less fit than those in the first fifteen of the Old Christians. Eduardo Strauch, strong at the outset, had been weakened by the revulsion he felt for eating human flesh immediately after the avalanche. The choice narrowed to Parrado, Canessa, Harley, Páez, Turcatti, Vizintín, and Fito Strauch. Some of them were more enthusiastic candidates than others. Parrado was so determined to escape that, had he not been chosen, he would have gone on his own. Turcatti too was emphatic that he should be one of the expeditionaries; he had two previous expeditions to prove his physical and mental stamina, and the younger boys had great faith that if he went the expedition would succeed.

Canessa had more imagination than some of the others and foresaw the danger and hardship which would be involved, but he felt that because of his exceptional strength and acknowledged inventiveness it was his duty to go. In the same way Fito Strauch volunteered, more from a sense of obligation than a real desire to leave the relative safety of the Fairchild, but nature intervened to settle his case, for eight days after the avalanche he developed severe hemorrhoids which effectively excluded him. His two cousins were delighted that he was to stay.

The remaining three, Páez, Harley, and Vizintín, all wanted to be expeditionaries but, though they were considered fit enough, some doubts were felt as to their maturity and strength of mind. And so it was decided that these three should go on a trial expedition which would last a day. Already, since the avalanche, there had been some minor sorties from the immediate surroundings of the plane. Francois and Inciarte had climbed three hundred feet up the mountain, resting after every ten steps to smoke a cigarette. Turcatti had gone up to the wing with Algorta, climbing with less energy and more effort than he had shown before, for he too had been weakened by his distaste for raw meat.

Páez, Harley, and Vizintín set out at eleven o'clock in the morning seven days after the avalanche to prove themselves. Their plan was to walk down across the valley to the large mountain on the other side. It seemed to be an attainable objective for a one-day expedition.

They wore two sweaters each, two pairs of trousers, and rugby boots. The surface of the snow was frozen so they walked easily down the valley, zigzagging where the descent was too steep to follow a direct path. They carried nothing with them to hamper their progress. After walking like this for an hour and a half they came upon the rear door of the plane and, scattered beyond it, some of the contents of the galley: two empty aluminum containers for storing coffee and Coca-Cola, a trash can, and a jar of instant coffee, empty but for a residue of powder left at the bottom. The three immediately put snow into the jar, melted it as best they could, and drank the coffee-flavored water. They then emptied out the rubbish bin and to their delight found some broken pieces of candy, which they scrupulously divided into thirds and sucked, sitting on the snow.

They were, for those few moments, in ecstasy. Though searching further, all they could find was a cylinder of gas, a broken thermos, and some maté. They put the maté in the thermos and took it with them as they continued on their way.

After walking down and across the valley for another two hours they began to realize that distances are deceptive in the snow and they were little nearer to the mountain opposite than when they started. Their progress had also become more difficult because the midday sun had melted the surface of the snow, and they now fell into it up to their knees. At three o'clock they decided to return to the plane, but as they retraced their tracks they quickly discovered how much more difficult it was to walk up the mountain than it had been to come down. Ominously, the sky had clouded over and a few flakes of snow began to fall and swirl around them in the wind.

They reached the coffee jar and refreshed themselves again with coffee-flavored water. Roy and Carlitos picked up the two containers from the galley, realizing that they would be useful for making water back at the plane, but found them too heavy and discarded them. Vizintín, however, held on to the large trash can and used it as a kind of staff to push himself up the mountain.

The climb became exceptionally difficult. They still sank to their knees in the snow, the slopes were steeper, the flurries turned to heavy snow, and all three were tired. Roy and Carlitos were close to panic. In the confused dimensions of the snow-covered landscape, they had lost all sense of how near or far they were from the plane. There were undulations in the side of the mountain, and as they reached the summit of each one they expected to see the Fairchild, but it was never there; and with each disappointment, their spirits fell. Roy began to cry, and Carlitos finally collapsed in the snow. "I can't go on," he said. "I can't, I can't. Leave me. You go on. Leave me here to die."

"Come on, Carlitos," said Roy through his tears. "For God's sake, come on! Think of your family . . . your mother . . . your father . . ."

"I can't, I can't move . . ."

"Get up, you sissy," said Vizintín. "We'll all freeze if we stay here."

"All right, I'm a sissy. A coward. I admit it. You go on."

But they would not leave, and they bombarded Carlitos with a mixture of exhortation and abuse that eventually brought him to his feet again. They climbed a bit farther, to the crest of another hill, and still the plane was not in sight.

"How much farther is it?" asked Carlitos. "How much farther?"

A little later he again collapsed in the snow.

"You go on," he said, "I'll follow you in a minute."

But again Vizintín and Harley would not abandon him, and once again they insulted him and pleaded with him until he got to his feet and walked on through the blinding snow.

They got back to the plane after the sun had set. The other boys had gone in and were waiting for them anxiously. When the three tumbled down the tunnel into the Fairchild, utterly exhausted, Carlitos and Roy in tears, it was apparent to all that the test had been severe and that some had failed.

"It was impossible," said Carlitos. "It was impossible and I collapsed, wanted to die, and cried like a baby."

Roy shivered, wept, and said nothing.

Vizintín's small, close-set eyes were quite dry. "It was tough," he said, "but possible."

Thus Vizintín became the fourth expeditionary. Carlitos withdrew his candidacy after his experience on the trial expedition, and Roy was told by Parrado that he could not be an expeditionary because he cried too much, whereupon Roy burst into tears. He was disappointed, however, only because he thought Fito was going. He had known Fito since they were children and felt safe by his side. When Fito developed hemorrhoids and withdrew, Roy was quite happy to be among those who were to stay behind.

Once the four expeditionaries had been chosen, they became a warrior class whose special obligations entitled them to special privileges. They were allowed anything which might improve their condition in body or mind. They ate more meat than the others and chose which pieces they preferred. They slept where, how, and for as long as they liked. They were no longer expected to share the everyday work of cutting meat and cleaning the plane, though Parrado and, to a lesser extent, Canessa continued to do so. And just as their bodies were coddled, so were their minds. Prayers were said at night for their health and well-being, and all conversation in their hearing was of an optimistic nature. If

Methol thought that the plane was in the middle of the Andes, he would make sure not to say so to an expeditionary. If ever their position was discussed with them, Chile was only a mile or two away on the other side of the mountain.

It was inevitable, perhaps, that the four should to some extent take advantage of their favored position and that this should provoke resentment. Sabella had to sacrifice his second pair of trousers to Canessa; Francois had only one pair of socks while Vizintín had six. Pieces of fat which had been carefully scavenged from the snow by some hungry boy would be requisitioned by Canessa, saying, "I need it to build up my strength, and if I don't build up my strength you'll never get out of here." Parrado took no advantage of his position, however; nor did Turcatti. Both worked as hard as they had done before and showed the same calm, affection, and optimism.

The expeditionaries were not the leaders of the group but a caste apart, separated from the others by their privileges and preoccupations. They might have evolved into an oligarchy had not their powers been checked by the triumvirate of the Strauch cousins. Of all the subgroups of friends and relatives that had existed before the avalanche, theirs was the only one to survive intact. The gang of younger boys had lost Nicolich and Storm; Canessa had lost Maspons; Nogueira had lost Platero; Methol had lost his wife. Gone too was Marcelo, the leader they had inherited from the outside world.

The closeness of the relationship between Fito Strauch, Eduardo Strauch, and Daniel Fernández gave them an immediate advantage over all the others in withstanding not the physical but the mental suffering caused by their isolation in the mountains. They also possessed those qualities of realism and practicality which were of much more use in their brutal predicament than the eloquence of Pancho Delgado or the gentle nature of Coche Inciarte. The reputation which they had gained, especially Fito, in the first week for facing up to unpalatable facts and making unpleasant decisions had won the respect of those whose lives had thereby been saved. Fito, who was the youngest of the three, was the most respected, not just for his judicious opinions but for the way in which he had supervised the rescue of those trapped in the avalanche at the moment of greatest hysteria. His realism, together with his strong faith in their ultimate salvation, led many of the boys to pin their hopes on him, and Carlitos and Roy suggested that he be made leader in place of Marcelo. But Fito refused this crown they offered him. There was no need to institutionalize the influence of the Strauch cousins.

Of all the work that had to be done, cutting meat off the bodies of their dead friends was the most difficult and unpleasant, and this was done by Fito, Eduardo, and Daniel Fernández. It was a ghastly task which even those as tough as Parrado or Vizintín could not bring themselves to perform. The corpses had first to be dug out of the snow, then thawed in the sun. The cold preserved them just as they had been at the moment of death. If the eyes remained open, they would close them, for it was hard to cut into a friend under his glassy gaze, however sure they were that the soul had long since departed.

The Strauches and Fernández, often helped by Zerbino, would cut large pieces of meat from the body; these would then be passed to another team, which would divide the chunks into smaller pieces with razor blades. This work was not so unpopular, for once the meat was separated from the bodies it was easier to forget what it was.

The meat was strictly rationed, and this again was done by the two Strauches and Daniel Fernández. The basic ration which was given out at midday was a small handful, perhaps half a pound, but it was agreed that those who worked could have more, because they used up energy through their exertions, and that the expeditionaries could have almost as much as they liked. One corpse was always finished before another was started.

They had, from necessity, come to eat almost every part of the body. Canessa knew that the liver contained the reserve of vitamins; for that reason he ate it himself and encouraged others to do so until it was set aside for the expeditionaries. Having overcome their revulsion against eating the liver, it was easier to move on to the heart, kidneys, and intestines. It was less extraordinary for them to do this than it might have been for a European or a North American, because it was common in Uruguay to eat the intestines and the lymphatic glands of a steer at an *asado*. The sheets of fat which had been cut from the body were dried in the sun until a crust formed, and then they were eaten by everyone. It was a source of energy and, though not as popular as the meat, was outside the rationing, as were the odd pieces of earlier carcasses which had been left around in the snow and could be scavenged by anyone. This helped fill the stomachs of those who were hungry, for it was only the expeditionaries who ever ate their fill of meat. The others felt a continuous craving for more, yet realized how important it was that what they had should be rationed. Only the lungs, the skin, the head, and the genitals of the corpses were thrown aside.

These were the rules, but there arose outside the rules an unofficial system of pilfering tolerated by the Strauches. This was why the task of cutting up the larger pieces was so popular; every now and then a sliver could be popped into one's mouth. Everyone who worked at cutting up the meat did it, even Fernández and the Strauches, and no one said anything so long as it did not go too far. One piece in the mouth for every ten cut up for the others was more or less normal. Mangino sometimes brought the proportion down to one for every five or six and Páez to one for three, but they would not hide what they did and desisted when the others shouted at them.

This system, like a good constitution, was fair in theory and flexible enough to allow for the weakness of human nature, but the burden fell on those who either could not or would not work. Echavarren and Nogueira were trapped in the plane by their broken, swollen, septic, and gangrenous legs, and only occasionally could they drag themselves down from their hammock and crawl out to defecate or to melt snow for drinking water. There was no question of their cutting up meat or scavenging in the snow. Delgado, too, had a broken leg, and Inciarte's leg was septic. Methol was still hampered by altitude sickness. Bobby Francois and Roy Harley were also crippled to some extent, not in their limbs but in their will; they could have worked but the shock of the crash or, in Roy's case, the shock of the avalanche followed by the abortive trial expedition seemed to have destroyed all sense of purpose. They simply sat in the sun.

The workers felt little compassion for those they thought of as parasites. In such extreme conditions lethargy seemed criminal. Vizintín thought that those who did not work should be given nothing to eat until they did. The others realized they had to keep their companions alive but saw no reason to do much more. They were harsh, too, in their assessment of the malingerers' condition. Some thought that Nogueira's legs were not broken and that he only imagined the pain he felt. They also thought that Delgado exaggerated the pain of his fractured femur. Mangino, after all, had broken his leg too, yet he managed to work at cutting up meat. They had little respect for Methol's altitude sickness or Francois's frozen feet. The result was that the only supplement to the ration for the "parasites" were the cells of their own bodies.

Some of the boys continued to find it difficult to eat raw human flesh. While the others extended the limit of what they could stomach to the liver, heart, kidneys, and intestines of the dead, Inciarte, Harley, and Turcatti still balked at the red meat of the muscles. The only occasions on which they found it easy to eat was when the meat was cooked; and every morning Inciarte would look across to Páez, who was in charge of this department, and ask, "Carlitos, are we cooking today?"

Carlitos would reply, "I don't know; it depends on the wind," for they could only light a fire if the weather was fair. But there were other factors involved. The supply of wood was limited; when they had used up all the Coca-Cola crates, there were only thin strips of wood which made up part of the wall of the plane. There was also Canessa's argument that proteins died at a high temperature and Fernández's that frying the meat made it shrink so that there was less to be eaten. Thus, cooking was allowed once or twice a week as the weather permitted, and on those occasions the less fastidious would hold back so that the others could eat more.

3

During the ten days between their choice of the four expeditionaries and November 15, when they hoped the cold weather would come to an end, the nineteen survivors developed both as a group and as individuals.

Parrado, for example, who before the accident had been a gawky, timid, would-be playboy, was now a hero. His courage, strength, and unselfishness made him the best loved of them all. He was always the most determined to brave the mountains and the cold and set out for civilization; and for this reason those who were younger, weaker, or had less determination placed all their faith in him. He also comforted them when they cried and took on himself much of the humdrum work around the plane from which, as an expeditionary, he was officially excused. He would never suggest a course of action without rising at the same time to put it into effect. One night, when part of the wall blew down in a strong wind, it was Parrado who climbed out from under the blankets to build it up again. When he returned he was so cold that those sleeping on either side of him had to punch and massage his body to bring back the circulation; but when, half an hour later, the wall blew down again, Parrado once again rose to rebuild it.

He only showed two areas of weakness. The first lay in his determination to get out. If he had had his own way he would have left straight after the avalanche with no proper preparation. He was patient with others but impatient with circumstances; he was unable to make the kind of detached assessment of the situation that came from Fito Strauch. If they had let him leave when he wanted to, he would not have survived.

His other failing was the irritation aroused in him by Roy Harley. It exasperated him that someone who was physically fit and strong should always be in tears; and yet others who were brought down in a similar way by the misery of their situation found Parrado their chief source of comfort. He was simple, warm, fair-minded, optimistic, and good-tempered. He rarely, if ever, swore and was most popular as a sleeping companion.

Next to Parrado, Numa Turcatti was the most generally beloved of the boys. He had a small, muscular body which from the first he had put to the service of their common cause. The expeditions prior to the avalanche had weakened him, and when Algorta had walked with him up to the wing he noticed that Numa no longer showed the same vigor. Also, his aversion to raw meat continued. Since he had known few of the boys before leaving Montevideo, it was a proof of his strength, simplicity, and complete lack of malice that he became so loved and respected by them. They all felt that if he and Parrado undertook an expedition it would succeed.

The other two expeditionaries did not inspire the same affection. It was recognized that Canessa had had good ideas, such as making the blankets and the hammocks which had done much to improve the conditions inside the plane. He knew about proteins and vitamins and had been a strong advocate of eating the flesh of the dead bodies. On the other hand his medical reputation, which had been so high at the time of his operation on Platero, had suffered after he had lanced one of the boils on Inciarte's leg and the infection had worsened as a result.

It was Canessa's personality, however, which made him difficult to live with. He was nervous and tense, bursting into bad temper at the smallest provocation, screaming imprecations and abuse in his high-pitched voice. Spasmodically brave and unselfish, he was more often impatient and stubborn. His nickname, Muscles, had been given him not just because of his physical strength but for this stubbornness of mind. On the rugby field it meant idiosyncratic play; in the plane it meant that he walked all over the sleeping bodies of his companions, pushing himself in wherever he chose. He did what he liked and no one could stop him. Only Parrado had some influence over him. The Strauches might have exercised some control, but they did not want to antagonize an expeditionary.

Vizintín was not as assertive and domineering as Canessa, but he was even more self-centered and did not have the compensating qualities of inventiveness and ingenuity. He had courage, as he had demonstrat on the trial expedition, but back in the plane his behavior was spoiled and childish. He quarreled with everyone, especially Inciarte and Algorta, and the only work he did was to melt snow for himself and do a few odd jobs that interested him; he made mittens for all the expeditionaries out of seat covers, and he made several pairs of sunglasses. At night he cried for his mother.

Only Canessa had any kind of control over Vizintín, but Mangino liked him. It was as if the three touchiest and most aggressive boys—each nineteen—had formed a small union. Mangino felt isolated; he too, like Turcatti, had not known many of the boys before and felt no restraint now at telling them to go to hell. In the days just following the accident he had been selfish and hysterical, but he came to work more for the group than most, in spite of his broken leg, and some of the other boys—notably Canessa and Eduardo Strauch—felt protective toward him.

Bobby Francois was another case where youth was deemed to excuse his shortcomings, the foremost of which was his lethargy. It was as if he had been born without an instinct for self-preservation; from the moment of the accident when he had sat down in the snow and lit a cigarette with the nonchalant remark, "We've had it," he had behaved as if survival was not worth the effort. He had been a lazy boy before they had left—his nickname was Fatty—but here sloth was suicidal, and had he been left to himself he would certainly have died. [illegible] not work. He would sit in the sun and melt snow [illegible] when he was forced to; otherwise he sat rub[illegible] which had been badly frozen in the ava[illegible]ght, when his blanket fell away from his body, he would not summon up the energy to cover himself again; someone else would do it. At one point Daniel Fernández had to massage his feet to stop them from becoming gangrenous.

There came a time when the cousins were so infuriated by Bobby's lethargy that they thought they would force him to work. They therefore told him that if he did not work he would get nothing to eat. Bobby merely shrugged his shoulders, looked mournfully up at them with his lovely large eyes, and said, "Yes, that's fair enough." But that morning he did as little as before, and when it came to midday and the distribution of food he did not take his plate and join the line. He seemed indifferent as to whether he lived or died and quite satisfied that the others should make the decision for him. They were not prepared to do so. Their "incentive" had failed. Bobby got his ration anyway.

Of the boys who were older and strong, Eduardo Strauch was, like Parrado, kind to those who were young and weak: to Mangino, Francois, and Moncho Sabella. Though uninjured, Moncho was weaker than most of the others and had a nervous nature. He had done well at the time of the accident—Nicolich considered that he had saved his life—and would have liked to prove himself the equal of all in courage and hard work, but he did not have the strength. He became one of the chorus who sat back in the sun, smoking, chatting, and melting snow, while others took the center of the stage.

Javier Methol, too, was one of the chorus. He was always dazed from the altitude sickness which continued to affect him. He talked a lot but stuttered and never finished his sentences. The boys, who were all at least ten years younger than he was, tended to regard him as a figure of fun. They called him Dumbo, because he had told them that this had been his nickname as a child, and laughed at him as he walked ponderously over the snow. They played practical jokes on him, in which he joined, exaggerating his condition because he knew it amused the boys and kept up their spirits. For example, one of their number would pretend that he had never eaten a cream bun (*ensaimada*), whereupon Methol would embark on a lengthy and pedantic description.

Just as he was reaching the end, a second boy would come up to him and say, "What are you talking about, Dumbo?"

"I'm describing a cream bun."

"A cream bun? What's that?"

"You don't know? Well, it's round, about this size . . ." and he would embark on a second description. Just as he reached the end of that a third boy would arrive and also pretend that he had never eaten a cream bun.

Methol's specialty was scraping oil off pieces of fat and collecting it as a laxative. He also was responsible for sharpening the knives, either on other knives or on pieces of rock. He made sunglasses, first for Canessa and then for himself, from plastic materials that were salvaged from the pilot's cabin. It was noticed, when he was doing this, that he only cut one lens from the plastic. This was the first intimation the boys had that he could only see out of one eye.

He would comfort the boys when they were unhappy. So too would Coche Inciarte, who was parallel with Parrado and Turcatti in the affections of the group. Parrado and Turcatti, however, were expeditionaries and so somewhat aloof, whereas Coche could understand their weaknesses because he was weak himself. He had done some work until his leg became infected, but later did nothing. He did not mind the smaller ration, because he did not like the meat when it was raw. He never grappled with their critical situation but spent the day dreaming of his life in Montevideo. Though the others were often exasperated by his inactivity, they all liked him too much to be angry. He had a most open and honest personality: kind, gentle, soft-spoken, and witty. No one would avoid his candid, smiling eyes, even if the object of their glance was to bum a cigarette or an extra piece of meat.

Pancho Delgado, on the other hand, though no more of a parasite than Inciarte, did not have the advantage of Inciarte's candid personality or his long-standing friendship with Fito Strauch. He was a boy of considerable eloquence and charm and had hitherto done well in life by the use of these talents. The Sartoris, for example, who had once been so opposed to his engagement to their daughter, had been won over by the bunches of flowers and the presents that he always brought when he came to call.

There were no flowers on the mountain, and charm and eloquence were not the qualities which were looked for in extreme conditions. Indeed, the eloquence that Delgado had shown now counted against him. The little mob did not forgive him his facile optimism. He was one of the oldest and should have known better than to raise their hopes without good cause. Thus, when he told them that he could not work because of his leg, some would not believe him and dismissed him as a malingerer.

It was a dangerous mood. A group in stress looks for a scapegoat, and Delgado was a likely candidate. His only friend from before was Numa Turcatti, who was too noble to know what was going on. All the others who did not pull their weight were somehow protected: Methol by his condition; Mangino, Sabella, Harley, and Francois by their youth; Inciarte by his good nature. Moreover, Inciarte did not pretend to be anything but what he was, whereas Delgado already had a lawyer's mind. He played life as if it was poker but did not realize that at that moment he had a weak hand. It was weak because he was hungry and he, almost alone among the nineteen, could neither pilfer himself nor count on friends and protectors to pilfer for him. It was a situation which was to become worse.

The effects of the trial expedition on Roy Harley and Carlitos Páez were contrary to what might have been expected. Roy, whose performance had been better than that of Carlitos, went into a decline. His rejection as an expeditionary led him to feel that he had disappointed his companions, and this, following so quickly on the death of his friend Nicolich, was as crippling to his mind as a broken bone would have been to his body. He became fragile, weeping if anyone spoke sharply to him and speaking in a high-pitched whine like a petulant child. He was lazy and selfish and could only be induced to do anything at all by oaths and abuse.

Carlitos, on the other hand, went the other way. The spoiled sissy and self-confessed coward became increasingly hard-working and responsible. He not only helped cut the meat but took on the duties of closing the entrance at night.

He had adverse qualities. He was bossy and quarrelsome and pilfered more than anyone else, yet his novel personality made a unique contribution to the morale of the group. Although the youngest, he was thick-set with a gruff voice like some giant teddy bear. His thinking was naïve, his pronouncements pompous, and his behavior often irresponsible—he would lose lighters and knives in the snow—yet up on the mountain, as in Montevideo, the mere thought of Carlitos brought a smile to the lips. He made them smile not so much because his jokes were funny but because of the comic effect of his whole personality. It was an important talent to possess, for there was little else to amuse them.

Carlitos Páez belonged to the second echelon of power. With Algorta and Zerbino, he acted as an auxiliary to the Strauch cousins. These three were the noncommissioned officers who received orders from those above and gave them to those below. Gustavo Zerbino, in particular, tended to flatter the older boys and bully the younger, though he was himself, at nineteen, one of the youngest in the group. He was affectionate but tense. Like Canessa, he became easily overexcited and would fly into a hysterical rage if, for instance, someone took his place in the plane opposite Daniel Fernández. It was to Fernández that he particularly attached himself. If Fernández asked for one of his pairs of trousers, he would give them to him; if Vizintín, an expeditionary, did the same, Zerbino would say, "Go to hell, you dirty brute. Find a pair for yourself."

With Fernández, Zerbino made himself responsible for collecting and guarding all the money and documents of those who had died. He also took it upon himself to investigate any misdemeanors such as moving in the night. For this, he was sometimes called "the detective." Before the accident, his nickname had been Ears, but this was changed to Caruso when it emerged during a conversation about food that Zerbino had never eaten *cappelletti alla Caruso* (a kind of ravioli with a sauce named after the singer Caruso) and did not even know what they were. His warm, simple nature made him easy to tease. Sometimes the boys would laugh at him because, in the late evening or early morning, he could not tell whether the sun was the moon or the moon the sun. He was also consistently pessimistic. If Fito sent him out to see what the weather was like, he would always return and say, "It's bitterly cold and there's a blizzard just getting under way."

Then Fito would turn to Carlitos and say, "You go and have a look." And Carlitos, who was an optimist, would return and report. "There's a little snow but it won't last long. In half an hour we'll have a clear blue sky."

Pedro Algorta was an unlikely hero. Following the criteria which brought down the spirit of some of the others, he should have been the first to go. Though he had started his schooling with the others at the Stella Maris College, he had continued it in Santiago and Buenos Aires (because of his father's work) and so knew few of the nineteen. Of his two friends, one, Felipe Maquirriain, was dead and the other, Arturo Nogueira, crippled and morose inside the plane.

There were several qualities which might have separated Algorta from the rest. He was shy, introspective, and a socialist, while they were boisterous, extroverted, and conservative. In Uruguay he had worked for the Frente Amplio, a kind of Popular Front which had presented itself to the electorate for the first time in the recent presidential election. Daniel Fernández and Fito Strauch, on the other hand, both belonged to the University National Movement (MUN), which supported Wilson Ferreira (a liberal Blanco); Eduardo Strauch supported Jorge Batlle (a liberal Colorado); while Carlitos Páez had voted for the reactionary Blanco, General Aguerrondo.

Another disadvantage for Algorta was his amnesia. He still could not remember what had happened in the days preceding the accident. He once ran around the plane with joy when Inciarte told him that an Argentinian team had won the football championships. It was quite untrue. More seriously, Algorta had entirely forgotten that his interest in going to Chile had been not just the cheap economics textbooks, or a first-hand study of South American socialism, but a girl whom he had met when he lived in Santiago. He had thought then that he loved her, but they had not met for a year and a half, and the letters they had exchanged had not been enough to sustain their feelings at the pitch he desired. His aim, then, had been to settle their relationship one way or the other, but now he had forgotten that it existed at all. Indeed, one of the reasons he wanted to get back to Montevideo was to find himself a girlfriend.

His mind was sufficiently alert for him to realize that he must work to survive, and his efforts gained the approbation of the cousins, especially Fito. He continued to feel slightly excluded, however—chiefly because he could not join in the conversation, which was so often about agriculture—but not so much as to develop a serious sense of isolation.

The three who were the government of this little community, Eduardo and Fito Strauch and Daniel Fernández, were not as individuals so different from the others. They dominated the group by virtue of the strength they brought to one another.

Daniel Fernández, for instance, was the oldest of the survivors, after Methol, and was conscious of the responsibility this placed on him. He was mature even for his age (he was twenty-six years old) and worked hard at keeping the cabin tidy, collecting the documents, and controlling the distribution of lighters and knives. He massaged Bobby Francois's frozen feet (for which Bobby promised to be his slave when they returned to Montevideo) and warned Canessa not to operate on Coche's infected leg. Though temperamentally shy, Daniel liked to talk and tell stories. He was calm, responsible, and fair. Indeed, the only qualities he lacked were physical strength and an assertive personality.

Eduardo Strauch, although nicknamed "the German," was in most ways less German than his two cousins. In appearance he took after his mother, an Urioste, having a smaller frame than Fito. His demeanor was attractive and his manner personable. He was the most urbane of the nineteen—perhaps because he had traveled in Europe—and had the most open mind. In general he was calm, but he was capable of passionate anger. He was inclined to be bossy, especially with Páez, but like Fito he was kind to the younger, more irritating boys, such as Mangino and Francois.

Fito Strauch was more temperamental than Eduardo, but he inspired more confidence in the group. When they considered their predicament, his thinking was always the most positive, his judgment the most sound. He also had to his credit the invention of the sunglasses which they needed to protect their eyes from the snow. Taking the sun visors from the pilot cabin, which were made of shaded plastic, he cut out two small circles and sewed them into plastic surrounds which had been cut from the cover of a folder containing the flight plan.

Fito was not free from failings. Like Daniel Fernández he was irritated by Mangino. He also quarreled with Eduardo as they settled down to sleep, and once he became so incensed at Algorta for lolling over onto him at night that he leaped to his feet and screamed at him, "You're killing me, you're killing me!"

Algorta just opened his eyes and said, "Oh, Fito, how can you?" and then went back to sleep.

4

The system which was evolved worked well. As in the Constitution of the United States, there were checks and balances. The Strauch cousins with their auxiliaries limited the power of the expeditionaries, and the expeditionaries limited the power of the Strauches. Both groups respected one another, and both acted with the tacit consent of all nineteen.

There were two among them who could play no part in the group because of the injuries they had suffered at the time of the accident. These were Rafael Echavarren and Arturo Nogueira. They both slept on the hammock that had been made by Canessa and rarely left the plane. It hurt them too much to walk, and to drag themselves out into the snow took almost more strength than was left in their limbs.

The two were quite different in background and temperament. Nogueira at twenty-one was a left-wing student of economics; Echavarren at twenty-two, a conservative dairy farmer. Their values were divergent and not likely to be reconciled on their bed of suffering, for at night the slightest inadvertent movement by one would cause great pain for the other.

Echavarren, a Basque in origin, had an open and courageous nature. The condition of his leg was appalling. The calf muscle which had been torn away from the leg had been pushed back into place, but the wound had become septic. Worse still, he was unable to move his leg or wriggle his feet at night, so the toes first went purple and then black as they were attacked by frostbite. During the day he would ask others to try and restore the circulation. *"Patroncito,"* he would say to Daniel Fernández, "give my legs a massage, would you? They're so numb that I can't feel them any more." And when Daniel had completed this task, Rafael would say, "I promise you, Fernández, that if I get out of this I'll give you all the cheese you want for the rest of your life."

He was utterly determined to escape. Every morning he would say to himself, "I am Rafael Echavarren and I swear I shall return," and when someone suggested that he write a letter to his parents or his *novia* he replied, "No, I'll tell them all about it when I get back." This faith made him popular with the other boys, as did his openness and honesty. When people knocked against his leg he would curse them and then, a minute or two later, apologize. He made them laugh, too, by eating an empty candy box, and he entertained them by describing how he made cheese on his dairy farm.

His condition got worse. His leg became heavy with pus, and the black skin color of gangrenous flesh spread from his toes to his foot. One morning, in a voice as emphatic and optimistic as ever, he asked for the attention of the whole group and told them that he was going to die. They protested, but he was firm. He told them, he said, because he wanted those who survived to convey his last wishes to his family, which were that his motorcycle should go to his steward and his jeep to his *novia*. The boys protested again, and the next day this sense of doom had left him and he returned to being among the most optimistic.

Arturo Nogueira was in better physical condition than Echavarren, but his mental state was the worst of all. Even before the accident he had been a brittle, difficult person, closed and silent even in his own family. The only person who had been able to draw him out of this introversion had been his *novia*, Inés Lombardero. She herself had suffered greatly in life—one of her brothers had been drowned with two other boys when their canoe had capsized off the coast at Carrasco. Her solace was Arturo; he would kiss her openly in the street.

His only other passion was politics. His strong sense of justice made him a militant idealist—sometimes socialist, sometimes anarchist. He had more or less abandoned the Church of Rome in favor of Utopia. Like Zerbino, he had worked in the slums of Montevideo at the behest of the Jesuits, but now he believed in more sweeping solutions to the problems of poverty and oppression.

In the plane he lay alone, his wide green eyes staring out of his emaciated face, a small beard on his chin. There was a time when he had shown some interest in their plight—above all in their precise position in the Andes—and had taken upon himself the role of cartographer, but as the days passed his faith diminished and the maps were put aside. He remembered that he had had a premonition as a child that he would die at the age of twenty-one. He told Parrado that he knew he was going to die.

Worse than this despair was his isolation within the group. He was barbed and moody with the others, and no one, in those conditions, took the trouble to penetrate his unfriendly exterior. Pedro Algorta was his only close friend, but Pedro himself was in danger of isolation and in no position to integrate Arturo against his will.

His antagonism toward the others was largely political. He and Echavarren quarreled ostensibly over blankets and the position of their feet, but it was the underlying divergence in their views that made their quarrels so acrimonious. There was also an occasion when Páez was entertaining the others with stories of his father. He told them how Páez Vilaró had been to Africa with Gunther Sachs and how Gunther Sachs and Brigitte Bardot had been to stay with them at Punta Ballena.

"Hey, Arturo, what do you think about all this?" Canessa asked.

"I'm not interested; I'm a socialist," said Arturo.

"You're not a socialist, you're a fool," said Canessa. "Stop trying to seem so hard-boiled."

"You're all oligarchs and reactionaries," said Arturo bitterly, "and I don't want to live in a Uruguay imbued with the kind of materialistic values that you represent . . . especially you, Páez."

"I'm not going to listen to him," said Carlitos.

"You may be a socialist," Inciarte said, stuttering with indignation, "but you're also a human being, and that's what counts up here."

"Ignore them," Algorta said to Nogueira. "It's all so unimportant."

Nogueira lapsed into silence, and later he told Páez that he regretted what he had said.

During the day, even when the sun was shining, Nogueira would remain in the plane. He would catch water which dripped through a hole in the roof, or Algorta, Canessa, and Zerbino would bring him water from outside. They would talk to him about his family and try and persuade him to go outside the plane, because some of them suspected that the injuries to his legs were imaginary, but all that these friends did to raise his spirits was to no avail.

It was cold, dark, and wet in the plane. Those who

remained inside breathed only its dank air. Nogueira became weaker, and it was only after a week that it was realized that he had not been eating his ration of meat. After that Algorta would take it in to him and place the small pieces of flesh into his mouth—wet from the saliva which dribbled out of it.

Parrado and Fito Strauch at last realized that Nogueira's isolation would kill him. Parrado came in to talk to him. "Do you want to stay here?" he asked him—"stay here" was the euphemism they used for death.

"I know I will," Arturo replied.

"You won't," said Parrado. "I'll get you out of here for Inés's birthday. You'll see."

One night as they were settling down to sleep, Arturo asked if he could lead the rosary. They agreed that he should, and Páez handed him his beads. Arturo then spoke his intentions, praying to God for their families, their countries, their companions who were dead and those who were there. He spoke with such feeling in his voice that the other eighteen—some of whom thought of the rosary as another way of counting sheep—were struck with a new respect and affection for him. When he had finished the five decades they were all silent; only Arturo himself could be heard weeping softly on the hammock. Pedro looked up at him and asked him why he was crying. "Because I am so close to God," Arturo replied.

He had among his belongings the list of clothes he had made before packing his suitcase. He took this and wrote on the other side, in a hand much weaker than before, a letter to his parents and his *novia*.

In situations such as this, even reason cannot understand the infinite and absolute power of God over men. I have never suffered as I do now—physically and morally—though I have never believed in Him so much. Physically this is torture—day by day, night by night—with a broken leg and swollen ankle, and the ankle of the other leg equally swollen. Morally and physically because of your absence and my longing to see you . . . and embrace you in the same way as my beloved Mama and Papa, to whom I want to say that I was wrong in the way I behaved toward them. . . . Strength. Life is hard but it is worth living. Even suffering. Courage.

A day later Arturo became weaker still and feverish. Pedro Algorta went up onto the hammock to sleep beside him and provide some warmth. They talked together about his family, Inés, the exams they would take together, and the football games they had watched on television. His talk was disconnected, and later he became delirious.

"Look, here comes the milk cart. Here's the farmer with the milk. Quick, open the door!" He raved on about the milk cart, then an ice cream cart, then Inés and lunch with his family on a Sunday. He was shivering with a high fever. Suddenly he started up and attempted to climb down onto the bodies of the sleeping boys below. Pedro clung to him, but Arturo screamed that Páez and then Echavarren were trying to kill him. Pedro held him and then hit him so that he fell back onto the hammock. Later he took some Librium and

Valium that they had among their medical supplies and fed them to his friend.

Arturo remained half in a coma, half delirious, throughout the next day, and that night it was so cold that they brought him down from the hammock to sleep on the floor. He was quieter now and he slept in Pedro's arms. It was in this position that he died. Methol and Zerbino tried to revive him with artificial respiration, but Pedro knew it was no use. He wept, and the next day, before the body was dragged out into the snow, he took the jacket and overcoat which had belonged to Arturo for himself.

5

Nogueira's death came as a shock to them all. It destroyed the thesis that those who had survived the avalanche were destined to live. Escape became more urgent, and the boys became impatient for the expeditionaries to set out, but they were still trapped in the plane for days at a time by the bitterly cold winds and driving snow.

In the days following the avalanche they had slept in no special order; the first to come in at night could take the warmest places. Later they evolved a stricter system which was designed to be more fair. Daniel Fernández and Pancho Delgado would take the cushions down from the roof where they had been drying in the sun and lay them out on the floor of the cabin. Then, at about five thirty, when the sun had gone behind the mountain and it became suddenly cold, the boys would line up in the order in which they were to sleep. First went Inciarte (but without Páez, who was his sleeping partner); then Fito and Eduardo; then Daniel Fernández and Gustavo Zerbino (unless it was their turn to sleep by the door). After them the order was not so fixed. Canessa would sleep where he liked, and Parrado usually slept with him. Francois and Harley stuck to each other. Methol slept with Mangino, Algorta slept with Turcatti or Delgado, and Sabella with Vizintín. The last of these pairs to go in would be the one whose turn it was to sleep in the coldest place by the door, but last of all came Carlitos, who had been given the job of shutting up the entrance every night in exchange for his place (with Inciarte) in the warmest part of the plane.

He was their *tapiador* (wall builder); but his place by the pilots' cabin involved another duty, which was to empty the plastic mug they used as a chamberpot through a hole in the fuselage. It was a tedious job because the mug was often smaller than the bladder that had need of it and had to be passed back for a second or even a third time, but there was no larger receptacle which was suitable. It was also continuously in demand, because the boys were often confined in the plane for fifteen hours at a time. Most were good enough to urinate before coming in, and would make use of the mug if they felt the need at around nine o'clock when the moon went in and they tried to sleep, but there were some—notably Mangino—who invariably awoke at three or four in the morning and called on Carlitos for the pot. On one occasion this irritated Carlitos so much that he pretended he could not find it and Mangino had to stumble out of the plane into the cold; on another he bartered his services for an extra cigarette.

At one point they tried to make a second "depot" at the entrance to the plane but found that when the snow melted so did the urine, and it seeped back into the cabin. Nevertheless, it was difficult for those sleeping by the entrance to ask for the pot at night because it meant waking everyone else to have it passed down. Algorta once woke up with this need and felt this inhibition; he therefore decided to piss onto the wall of snow. The next morning, by the light of day, he saw that he had urinated all over someone's tray of grease. He said nothing.

The inside of the plane became a mess. It was not just the urine which soiled it but scraps of fat and bone which were left on the floor. After a time a new rule was made that no bones were to be brought into the plane and any fat that was brought in was to be taken out again the same day. All the same, the snow at either end remained filthy, and only the cold prevented an overpowering stench.

It was difficult to sleep. They were packed so tightly together that, if one moved, everyone else had to move and all the flimsy seat covers came off their bodies. They also suffered from an understandable fear of a second avalanche. They could always hear odd sounds outside the plane—either the rumbling of the Tinguiririca volcano or other avalanches elsewhere in the mountains. Stones would break loose from the mountains and tumble down toward them. One once hit the plane as they were trying to sleep, and Inciarte and Sabella leaped to their feet, thinking it was another avalanche.

The rest were always ready to do the same. Methol slept in a sitting position, his head covered by a rugby shirt to warm the air he breathed. Then, as he dropped off, he would loll forward or fall to one side, to the discomfort and annoyance of whoever slept beside him.

It was this kind of irritation which caused the only quarrels which led to a fight. They all swore at one another for sticking a foot in a face or stealing a blanket, but on only a few occasions did this lead to blows. Canessa and Vizintín were the most uncontrolled in this respect. They were stronger than the others and took advantage of their status as expeditionaries to sleep how and where they liked, though they were careful not to antagonize Parrado, Fernández, or the Strauches. On one occasion Vizintín lay down with one foot in Harley's face because Harley would not make room. When asked to move it, he would not. Roy then pushed the foot off his face, so Vizintín kicked him. At this Roy became furious and would have attacked Vizintín had not Daniel Fernández intervened. On another occasion Vizintín kicked Turcatti, and Numa, who normally had the sweetest temper, flew into a rage and screamed at him, "You dirty brute, for as long as I live I'll never talk to you again!" And Inciarte, taking Turcatti's side, joined in, "You son of a whore, get your leg out of the way or I'll smash your face in!" Vizintín told them both to go to hell, and once again Fernández intervened and told them all to calm down.

Inciarte also quarreled with Canessa, who raised his hand to hit him, but Inciarte said, "If you do that, I'll break your neck"—brave words for someone who was now among the weakest, but enough to make Canessa think better of what he was about to do. This quarrel, like most of the others, ended as quickly as it had started, with further tears, embraces, and repetition once again of the general view that if they did not stick together they would never get away.

The bickering, threatening, cursing, and complaints were also the only way in which they could release the intense frustration which had built up inside them. If someone knocked against Echavarren's leg, for example, he would scream out of all proportion to the pain it had caused, thereby giving vent to the nagging agony that he suffered all the time. In the same way it made many of them feel better to shout "big balls" at Vizintín or call Canessa a "son of a whore." What was strange was that some—especially Parrado—never quarreled at all.

One night Coche Inciarte dreamed that he was sleeping on the floor of his uncle's house in Buenos Aires. Mangino was sleeping next to him, rubbing up against his infected leg. In the dream Coche began to kick him; he then heard shouting and awoke to find Fito and Carlitos shaking his shoulders and Mangino in tears beside him. His dream had come true—except that he was not in his uncle's house in Buenos Aires but in the wreck of a Fairchild in the middle of the Andes.

6

Before sleeping at night they would talk together. Several subjects were touched upon, such as rugby, which most of them played, or agronomy, which most of them studied, but somehow they would always end up discussing food. What they lacked in their daily diet they made up for in their imagination, and when each boy had exhausted his own menu, he would prey on that of another. Echavarren, for instance, had a dairy farm and could tell them about cheese, lingering over every detail of its manufacture and describing the taste and texture of each different variety with such passion that many of the others were left wondering why they too were not dairy farmers.

To draw out the conversation, and dredge everyone's memory for the smallest scrap, they would categorize. Each boy would have to describe a dish that was cooked at home, then something he could cook himself. After that came the *novia*'s specialty, then the most exotic food he had eaten, then his favorite pudding, then a foreign dish, then something that was cooked in the countryside, then the oddest thing he had ever eaten.

Nogueira, before he died, presented them with cream, meringues, and *dulce-de-leche*, a thick sweet sauce made with milk and sugar, with a taste between that of condensed milk and caramel cream. Harley, as a winter dish, thought of peanuts and *dulce-de-leche* coated with chocolate, and, in summer, peanuts and *dulce-de-leche* ice cream. Algorta had no dish which he could cook himself, but he offered the boys the *paella* his father sometimes made and his uncle's *gnocchi*. Parrado could promise the *barenkis* cooked by his Ukrainian grandmother; and for those who did not know what they were he described these little pancakes filled with cheese, ham, and mashed potatoes. Vizintín, who always spent the summer by the sea near the border with

Brazil, described a bouillabaisse, and Methol told him that when they returned he would get Vizintín to show him how to make it.

Numa Turcatti was listening to this conversation. "Methol," he began.

"If you call me Methol I won't answer you."

Numa was formal as well as shy. "Javier," he said. "When you make that bouillabaisse, will you ask me along?"

"I will," said Methol, and he smiled, for though Numa had started the sentence with the second person singular, *tú*, he had retreated in the second half to the more formal *usted*.

Methol was their expert on food. The cream bun was by no means his only contribution. He had lived the longest and so eaten the most, and when they started to make a register of all the restaurants they knew in Montevideo, each with its specialty, he was the one who could name the most. Inciarte made the list in a notebook which had belonged to Nicolich, and when the last restaurant with its most obscure specialty (*cappelletti alla Caruso)* had been entered into it, they numbered ninety-eight.

Later they had a competition to see who could invent the best menu—including wines—but by then these imaginary feasts had come to cause more suffering than joy. It depressed them when they came out of their gourmandizing dreams to the reality of raw flesh and fat. They were also afraid that all the digestive juices that were released by their fantasies would give them ulcers. They therefore came to a tacit agreement that conversation about food should stop. Only Methol continued.

Alas, though they might consciously decide to cut food from their minds, they had no control over their dreams. Carlitos dreamed of an orange suspended in air above him. He reached for it but could not touch it. On another occasion he dreamed that a flying saucer came and hovered over the plane. Stairs were lowered and a stewardess came out. He asked her for a strawberry milkshake but was given only a glass of water with a strawberry floating on the top. He flew off in the flying saucer and landed at Kennedy Airport, New York, where his mother and grandmother were there to meet him. He crossed the lobby and bought a glass of strawberry milkshake but it was empty.

Roy dreamed that he was in a bakery where biscuits were being shoveled out of the oven. He tried to tell the baker that they were up in the Andes but could not make him understand.

Their inclination was to think and talk about their families. For this reason Carlitos liked to look at the moon; it consoled him to think that his mother and father would be looking at the same moon in Montevideo. It was one of the disadvantages of his place in the plane that he could not see out of a window, but once, in exchange for the pot, Fito held up a pocket mirror so that Carlitos could see the reflection of his beloved moon.

Eduardo would talk to Fito about his trip to Europe, or the two double cousins would discuss their families, but often when they did so they would hear the rhythmic sniffing which told them that Daniel Fernández had started to sob beside them. It was too painful to think of home, and to protect their sanity most boys succeeded in excluding such thoughts from their minds.

There was little else to talk about. Most were deeply interested in Uruguayan politics but were wary, after Nogueira's outburst, of discussing anything which might inflame conflicting passions. When they heard on the radio, for example, that the Colorado politician Jorge Batlle had been arrested for criticizing the Army, Daniel Fernández—a Blanco—leaped in the air with delight: yet both Canessa and Eduardo had voted for Batlle in the recent presidential election.

The safest topic of conversation was agriculture, because many were training or already working as farmers and ranchers or else had a farm or ranch in the family. Páez, Francois, and Sabella all had properties in the same part of the interior, and Inciarte and Echavarren both ran dairy farms.

Occasionally Pedro Algorta would feel excluded from the group because he knew nothing of country matters. Seeing this, the farmers would try and draw him in. They planned a Regional Consortium of Agricultural Experimentation in which Pedro was to have charge of the rabbits. They would all live together on some land that Carlitos owned in the Coronilla in adjoining houses designed by Eduardo.

They all thought a great deal about their Regional Consortium—especially Methol, who, with Parrado, was to run the restaurant. One evening, when they were lying in the plane waiting to go to sleep, Methol leaned across to Daniel Fernández and asked him if he would not mind moving away, as he wanted to ask Zerbino something personal. Fernández did as he was asked—disrupting the whole line of recumbent figures—whereupon Methol whispered in Zerbino's ear to ask whether he would do the accounts of the restaurant.

The Regional Consortium was an honest project but the restaurant was its weakness, and soon they talked less about methods for fattening cattle or improving grain and more about the plovers' eggs and suckling pig that would be served in the restaurant. It was hard to keep off food, too, when they planned the parties they would have with their *novias* on their return to Uruguay. They did not envisage asking anyone outside the group, and when they thought of their *novias,* or talked about them, it was always with purity and respect. They had too great a need of God to offend Him with salacious thoughts and conversation. Death was too close to risk even the smallest sin. Moreover, all sexual feeling seemed to have left them, due no doubt to the cold and their own debility. Some even became slightly alarmed that their inadequate diet would make them impotent.

There was therefore no sexual frustration in a physical sense, but there did exist a great emotional need to think of a partner in life. The letters written by Nogueira and Nicolich had been addressed more to their *novias* than to their parents. Those who still lived and had *novias*—Daniel Fernández, Coche Inciarte, Pancho Delgado, Rafael Echavarren, Roberto Canessa, Alvaro Mangino—thought of them intensely and with the greatest devotion. Pedro Algorta, as we have seen, had forgotten about the girl who was waiting for him in Santiago and was impatient to get back to Uruguay and find a *novia*. Zerbino was not engaged, but he talked to the others about a girl he had met, and by general acclaim she became his *novia*.

They were not induced by the extremity of their situation to talk at length about the more fundamental philosophical issues of life and death. Inciarte, Zerbino, and Algorta—who were the three most politically progressive among the eighteen who were still alive—once discussed the relationship between religious faith and political responsibility. On another occasion Pedro Algorta and Fito Strauch discussed the existence and nature of God. Pedro was well trained by the Jesuits in Santiago and could explain the philosophic theories of Marx and Teilhard de Chardin. Both he and Fito were skeptics; neither believed that God was the kind of being who watched over the destiny of each individual. To Pedro God was the love which existed between two human beings, or a group of human beings. Thus love was all important.

Carlitos tried to join in this conversation—he had his own notions about God—but Fito and Pedro told him that his mind was too slow to keep up with the argument. Carlitos had his revenge the next day when Pedro swore at someone for kicking his face or stepping on his tray of fat: "Oh, but why are you saying such dreadful things, Pedro? I thought love was everything?"

There was nothing to read but one or two comics. No one played games, sang songs, or told anecdotes any more. There was the odd coarse joke about Tito's piles, and they laughed when Inciarte stretched up to fetch something from the hat rack and brushed his face against a lifeless hand which had been brought in to stave off hunger in the night. There was an occasional witticism about eating human flesh—"When I go to the butcher in Montevideo, I'll ask to taste it first"—and the chances of their own death—"How would I look in a piece of ice?" They also invented words or changed the ending to words—especially Carlitos—and evolved phrases and slogans either to rally their morale or to express a stark truth which they could not bring themselves to express more clearly. "The loser stays" was the nearest they came to saying that the weak would die. "A man never dies who fights," they would say, or "We've beaten the cold," and over and over again they would repeat the only fact they knew to be true: "To the west is Chile."

For that, finally, was their chief preoccupation and topic of conversation—their escape. The expedition was planned over and over again. Its equipment was discussed, designed, and manufactured. The route was discussed by the whole group. There was never any doubt but that the expeditionaries were acting for them all and would follow the instructions of the majority. The more practical thought about how they could insulate the feet; the dreamers discussed what they would do when they got to Chile, deciding that they would telephone their parents in Montevideo to say that they were alive and then take the train to Mendoza. They thought that when they got back to Montevideo they might find a journalist who was interested in what they had been through, and they also planned to write a book for which Canessa chose the title "Maybe Tomorrow," because they always hoped that something encouraging would take place the next day. At around nine o'clock, when the moon had disappeared over the horizon, they would stop talking and get ready for sleep. Carlitos would start the rosary, making the same intentions every night—for his father, his mother, and the peace of the world. After that Inciarte or Fernández would say the second mystery, and Algorta, Zerbino, Sabella, Harley, or Delgado would share the rest. Most of them believed in God and their need of Him. They found great comfort, too, in praying to the Mother of God, as if she was in a better position to understand how much they longed to return to their families. They sometimes said the "Hail Holy Queen," thinking of themselves as the "poor banished children of Eve" and the valley in which they were trapped as the "vale of tears." They were always frightened of another avalanche, especially when a storm blew outside the plane, and one night, when the winds were particularly violent, they prayed a rosary to the Virgin to protest them—and by the time they had finished, the storm had died down.

Fito remained skeptical. He thought of the rosary as a sleeping pill—something which kept one's mind off depressing subjects and sent one to sleep by its monotony. The others knew of his attitude, and one night they made use of it. The ground beneath the plane had started to tremble with the twitching of the Tinguiririca volcano, and all their terror returned that this movement would disturb the huge quantities of snow above them and send down an avalanche that would bury them forever. They thrust the rosary into Fito's hands and told him to pray. The skeptic was as frightened as the believers. He said the rosary with the most specific intention that they might be saved from the volcano;

and by the time he had finished the decade the rumbling had stopped.

7

There were two further matters which continually preoccupied them. The first was cigarettes. Parrado, Canessa, and Vizintín were the only nonsmokers among them. Zerbino had not smoked before but had taken it up on the mountain. The rest were all heavy smokers, and because of the added stress of the conditions in which they were living they would all have liked to smoke even more than they were used to.

It so happened that there was no real shortage of cigarettes. Javier Methol and Pancho Abal, who had both worked for a tobacco company and knew of the shortage of tobacco in Chile, had come loaded with cartons of Uruguayan cigarettes.

All the same, there was rationing. One pack of twenty had to last each boy for two days, and most managed to exercise sufficient control over themselves to space the ten cigarettes through the day. The feckless, however—especially Inciarte and Delgado—would finish their pack on the first day and find themselves with nothing to smoke on the second. Their only chance in such a situation was either to get their future ration in advance or to scrounge cigarettes from the more provident. It was in these conditions that Delgado would remember, say, what a good friend he was of Sabella's brother or Inciarte would invite Algorta to an especially delicious dinner when they returned to Montevideo.

They would smoke their first cigarette of the day lying in the plane when they had just waked up. Then one boy would try and coax another out into the snow.

"It looks lovely, why don't you go out?"

"Why don't you?"

Eventually one would rise, find his shoes, rub them together to thaw them out, put them on, and then pull away the cases and clothes with which Carlitos had blocked the entrance the night before. Each boy took some cushions with him, when the sun was shining, to dry them out on the roof. They would dry themselves out, too, for they never changed their clothes or took them off but only added to them. The blankets would be piled onto the hammocks, and the last out of the cabin would have to tidy it up.

In the course of the morning the cousins would set to work cutting meat from a body while others would take advantage of the hard surface of the snow to scavenge for discarded gobbets of fat and offal or to go to a hole at the front of the plane and try to defecate.

This was their second great preoccupation, because the diet upon which they were living—raw meat, fat, and melted snow—gave rise to the most chronic constipation. Day after day, then week after week, would pass with nothing to show for the most strenuous efforts. Some began to fear that their intestines would split, and every method was used to facilitate the delivery of their feces. Zerbino used a small stick to prize them out and Methol—one of the worst sufferers—swallowed oil that he scraped off the fat as a laxative. Carlitos used the same oil to make a laxative soup (when cooking) for himself and for Fito who, with his hemorrhoids, had special need of it.

It was a wretched situation, not without its comic side. The boys began to place bets as to who would go last. There was an occasion when Moncho Sabella, crouched in the snow trying to defecate, said, "I can't do it, I can't do it."

Vizintín started to laugh at him—"You can't do it, you can't do it"—whereupon Sabella made an extra effort, succeeded, and threw the rock-hard result at his tormentor.

Javier Methol was one of the last. Day after day he sat over a cushion counting his money, waiting for his efforts to be rewarded, and when at last he succeeded, he announced his victory to the whole group and they applauded. That night, when he complained that he was uncomfortable, they all shouted him down. "Shut up," they said. "You crapped, so shut up."

The competition drew to a close. After twenty-eight days on the mountain, Páez managed to defecate; Delgado after thirty-two; and the last, Bobby Francois, after thirty-four.

Ironically, this acute constipation was followed by an epidemic of diarrhea. Their own diagnosis was that it came from eating too much fat, though it may well have resulted from their inadequate diet. Algorta never suffered from it and ascribed his immunity to the cartilage which he ate, in contrast to most of the others.

It was a miserable addition to their afflictions. One night Canessa was taken short and came out of the plane to find half a dozen other figures crouching in the moonlight. This scene particularly depressed him—he thought it was the end of everything—and, thereafter, though he continued to suffer from diarrhea, he never came out but defecated onto a blanket on a rugby shirt inside the plane. This infuriated the others, but Canessa was cussed and stubborn and there was nothing they could do about it. Carlitos was especially angry because once, by accident, he picked up a shirt to block up the entrance and found it covered with Canessa's ordure.

Sabella suffered the worst attack of diarrhea. It continued for some days and he got progressively weaker, until finally one night he grew delirious. The other boys became alarmed. Canessa advised him not to eat so much—and above all, not to eat any fat. Sabella, however, believed in consistency. He had always taken ten paces every day by way of exercise and felt that if he took one less it would be the start of an irreversible decline. For the same reason he thought it would be dangerous to deprive himself of food, but when the cousins saw that he was continuing to feed himself they cut off his ration and confined him to the plane.

The next day Sabella went out to defecate and returned with the news that his diarrhea was cured; but he had not taken account of Zerbino, who as doctor and detective had examined the evidence of this "cure." He denounced Sabella to the others, who sent him back into the plane without his ration of fat.

Much as he resented this treatment, it proved effective. He was cured of diarrhea and later regained something of his strength.

8

As it drew nearer to the fifteenth of November, an atmosphere of excitement and anticipation grew up in the plane. There were repeated discussions of who would be the first to telephone their parents and how casual and blasé they would be about their escape. They also lingered over the meat pastries they would buy in Mendoza on their way back. From there they would take a bus to Buenos Aires and then a boat across the River Plate. As they reached each stage of the journey in their minds, they pondered on what they would eat. They knew that Buenos Aires had some of the best restaurants in the world, and they hoped that by the time they were on the boat their stomachs would be full enough for them to stop eating and buy presents for their families.

The expeditionaries themselves were more preoccupied with the practical problems which faced them—especially protection against the cold. Each assembled three pairs of trousers, a T-shirt, two sweaters, and an overcoat. They had the three best pairs of dark glasses. Vizintín had those which belonged to the pilot; he also wore the pilot's flying helmet. Canessa constructed knapsacks out of trousers. He tied nylon straps to the end of each leg, brought them around his shoulders, and threaded them through the belt holes. Vizintín made six pairs of mittens from the seat covers.

They knew from earlier expeditions that the chief problem which would face them was the insulation of their feet against the cold. They had rugby boots, and Vizintín had prized from a reluctant Harley the stout shoes that Nicolich had been given by his *novia,* but they had no thick socks. Then they came up with the idea that they should provide their feet with an extra layer of fat and skin from the dead bodies outside. They found that if they made two incisions—one in the middle of the elbow, the other in the middle of the forearm—pulled away the skin with its subcutaneous layer of fat, and sewed up the lower end, they were left with a rudimentary pair of socks with the dead skin of the elbow fitting neatly over the live skin of the heel.*

The only setback they suffered as the date of their departure approached was that someone stepped on Turcatti's leg and the resulting bruise began to go septic. Numa, however, dismissed this as insignificant and at first no one was seriously concerned. Their minds were more on the route the expeditionaries were to take, for in assessing their position and therefore the direction they should go they were faced with two conflicting pieces of evidence. They knew from the dying words of the pilot that they had passed Curicó, that Curicó was in Chile, that Chile was to the west. They also knew, however, that all water flows to the sea; and the plane's compass, which was still intact, showed that the valley they were in ran down to the east.

The only answer that seemed to satisfy all the criteria was that the valley curved around the mountains to

---

*Compare this with the expedient of their common forefather, the South American gaucho. "His boot (the *bota de potro*) was the hide stripped from a colt's hind leg and pulled on to his own leg while still moist so that it dried to the appropriate shape, the upper part forming the boot's leg, the hock fitting over the heel, and the remainder covering the foot, with an aperture for the big toe." George Pendle, *Argentina* (London: Oxford University Press, 1955).

the northeast and doubled back on itself to run west. On this assumption the expeditionaries planned to set off down the valley, even though this would be walking away from Chile. The mountains behind them were so immense that there was no question of climbing over them. To go west they could only go east.

The boys awoke early on the morning of November 15 and helped the expeditionaries to put on their equipment. It was snowing outside, but by seven o'clock the four had set out. Parrado had taken one of the small red shoes he had bought for his nephew and left the other hanging in the plane, saying to the others that he would be back to fetch it. He was back sooner than they thought. The snow got much worse, and after three hours they returned.

There followed two days of weather as bad as any they had experienced, with a high wind and a blizzard blowing outside the plane. Pedro Algorta, who had told them all that summer set in on the fifteenth, became for a time the butt of their disappointment and hostility. And in those extra days they waited Turcatti's leg became worse. There were now two boils the size of hen's eggs, and Canessa lanced them both to remove the pus. It was extremely painful for Numa to walk on his septic leg, yet when Canessa told him that he was not fit to go on the expedition Numa became angry. He insisted that he was well enough, but it was clear to all that he would only hold them back, and he was obliged to accept the decision of the majority. On the morning of Friday, November 17, after five weeks on the mountain, they awoke to a clear blue sky. There was nothing now to stop the depleted force of expeditionaries. They filled their knapsacks with liver and meat (stuffed into rugby socks), a bottle of water, seat covers, and the traveling rug which Señora Parrado had brought with her on the plane.

The others all trooped out of the plane to see them off, and when Parrado, Canessa, and Vizintín had disappeared over the first horizon they began to place bets on when they would reach civilization. They were sure they would all be in Montevideo in three weeks' time because they had planned in detail the party they would have for Parrado's birthday on December 9 (including the dish each would bring), but they assumed their expeditionaries would reach Chile much sooner than that. Algorta thought it would be the following Tuesday; Turcatti and Francois on Wednesday. Six of them wagered on Thursday, from Mangino, who thought they would reach help at ten in the morning to Carlitos, who put it at half past three. Harley, Zerbino, and Fito Strauch bet on Friday; Echavarren and Methol on Saturday; and Moncho Sabella—the most pessimistic—estimated that they would reach civilization at twenty minutes past ten a week from Sunday.

9

Canessa led the expedition, pulling as a sled half a Samsonite suitcase on which were piled the four rugby socks filled with meat, the bottle of water, and the cushions they would use as snowshoes when the sun melted the hard surface of the snow. Vizintín came next, loaded like a packhorse with all the blankets, and Parrado brought up the rear.

They made quick progress toward the northeast. They were going downhill and their rugby boots gripped well on the frozen snow. As they progressed, Canessa drew ahead, and after walking for two hours Parrado and Vizintín heard him shout and then saw him wave at them. He had stopped at the top of a crest of snow, and as they caught up with him Canessa said, "I've got a surprise for you."

"What?" asked Parrado.

"The tail."

Parrado and Vizintín reached the top of the hillock of snow and there indeed, a hundred yards ahead of them, was the tail of the Fairchild. It had lost both its wings but the cone itself was intact. What immediately excited their interest were the suitcases that they could see scattered around it. They ran over, opened them, and rummaged through their contents. It was like finding treasure; there were jeans, sweaters, socks, and Panchito Abal's skiing kit. In Abal's suitcase they also found a box of chocolates, from which they immediately ate four each but decided to ration the rest.

The three boys then stripped off the filthy garments they were wearing and changed into the warmest clothes they could find. Canessa and Parrado took off the stockings made of human skin and threw them aside. There were now plenty of good woolen socks, and they took three pairs each. Vizintín took four to pad out Nicolich's boots, which were too big for him. He also took the balaclava which formed part of Abal's skiing equipment, and Parrado took the boots.

Next they went into the tail itself and found, in the galley, a packet of sugar and three Mendozan meat pastries. The latter they ate at once; the sugar they kept for later. Behind the galley there was a large, dark luggage compartment in which there were more suitcases. They opened them all, pulling out the clothes and scattering them on the floor. In one they found a bottle of rum, and in many there were cartons of cigarettes.

They searched for the plane's batteries which the mechanic, Roque, had told them were in the tail section and found them through a small hatch on the outside of the plane. They also found more Coca-Cola crates and comic books, with which they made a fire. Canessa began to fry some of the meat they had brought with them, while Vizintín and Parrado continued to rummage inside the tail. They found some sandwiches wrapped in plastic which were moldy, but they unwrapped them and salvaged what was edible. Then they ate the meat they had cooked and finished it off with a spoonful of sugar mixed with chlorophyll toothpaste in half an inch of rum. Never in their lives had a pudding tasted so delicious.

The sun went behind the mountains and it began to grow cold. Vizintín and Parrado brought in all the clothes from around the tail and scattered them over the floor of the luggage compartment while Canessa traced the wires which led from the batteries and attached them to a lightbulb he had taken from the galley. He connected it but the bulb burst. He tried another, and this time it lit up. The three then climbed into the luggage compartment, blocked up the door with suitcases and clothes, and lay back on the floor. Because of the light they could read comics before going to sleep. After the cramped conditions back at the plane, it was delightfully warm and comfortable. At nine Canessa disconnected the bulb. They had eaten well and now they slept soundly.

The neat morning it was snowing slightly but they loaded up the Samsonite sled, filled their knapsacks, and went on down the valley to the northeast. They could see an enormous mountain to their left and estimated that it might take them three days to walk around it to where the valley would turn to the west.

The snow stopped, the sky cleared, and toward eleven o'clock in the morning it began to get very hot. The sun beat down on their backs and was reflected up at their faces by the snow. Every now and then they would stop to strip off a pair of trousers or a sweater, but it took a lot of their energy to do this and the garments were almost as tiresome to carry as they were to wear.

At around midday they came to an outcrop of rock over which flowed a small trickle of water. It was almost a stream, and they decided to stop there and shelter from the sun, constructing a tent with the blankets and the metal poles they carried. They ate some of their meat and Vizintín went to drink some of the water, but it looked brackish and the other two preferred to make water from melted snow.

As they lay there in the shade they stared at the huge mountain ahead of them. Its size defied all calculations of its distance from where they were. As the light changed it seemed to move farther away and the distant shadow where the valley might turn to the west farther

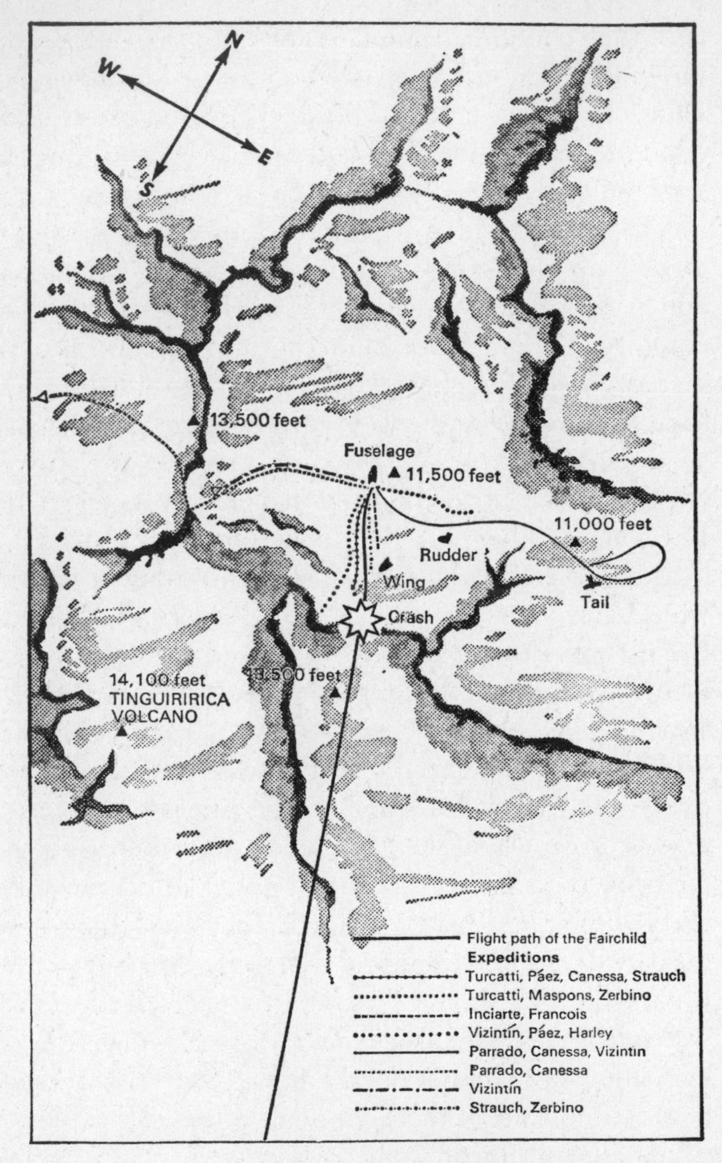

The area of the crash

still. The more Canessa studied what lay ahead, the more skeptical he became about their strategy. From what he could see, the valley continued to go east; thus every step they took, he thought, would take them farther into the Andes. But he said nothing about this to the other two that afternoon.

They were tired and the sun was hot, but no sooner had it left them to sink behind the mountains to the west than the temperature plummeted to freezing and the light began to fade. They therefore decided to spend the night where they were. They dug a hole in the snow to give themselves some protection and, once they were lying in it, covered themselves with the blankets they had brought with them.

It was a beautiful night with a clear sky. Because of the altitude they could see millions upon millions of bright stars. The air was quite still; there was no wind at all. Their situation might even have been enviable had it not been so cold, for as the night continued the temperature sank lower and lower and the three expeditionaries began to freeze. Their clothes and blankets seemed hardly to warm them at all. In desperation they lay on top of one another—Vizintín at the bottom, Parrado in the middle, and Canessa on the top. In this way they warmed one another with their bodies but got very little sleep.

Canessa and Parrado were both awake when the sun rose the next morning. "It's hopeless," said Canessa. "We won't survive another night like that."

Parrado stood up and looked toward the northeast. "We've got to go on," he said. "They're counting on us."

"We'll be no use to them lying dead in the snow."

"I'm going on."

"Look," said Canessa, pointing toward the mountain. "There isn't an opening. The valley doesn't go to the west. We're just walking farther into the Andes."

"You never know. If we go on—"

"Don't fool yourself."

Parrado looked again to the northeast and saw little there to encourage him. "Then what do you suggest we do?" he asked.

"Go back to the tail," said Canessa. "Take out the batteries and take them back up to the plane. Roque said that with the batteries we could make the radio work."

Parrado looked doubtful. He turned to Vizintín, who by this time had waked up. "What do you think, Tintin?"

"I don't know. I'll go along with whatever you two decide."

"But what do you think we should do? Should we go on?"

"Maybe."

"Or should we try and make the radio work."

"Yes. Perhaps we should do that."

"Which?"

"I don't mind."

Parrado became furious with Vizintín for his indecisiveness and tried to make him come down on one side or the other. Eventually Vizintín sided with Canessa, for, as Canessa said, "If we nearly froze to death on a clear night, think what would happen if there was a storm. It would be suicide."

They set off back toward the tail, and though it was considerably more difficult climbing up the valley than it had been coming down, they reached it by early afternoon and collapsed onto the great comfort of the clothes-strewn floor of the luggage compartment. It was a most luxurious habitation, sheltering them from the sun during the day and keeping out the cold at night, and they were tempted by its comfort into staying there throughout the two following days. By then, however, their supplies of meat were running low, so they decided to return to the Fairchild. Canessa and Vizintín climbed through the small hatch into the part of the plane where the batteries were stored, disconnected them, and handed them out to Parrado. Vizintín also found that the large tubes which were part of the plane's heating system were wrapped around with an insulating material about two feet wide and half an inch thick made of plastic and some artificial fiber. He cut off some strips of it, thinking it might make a good lining for his jacket.

The batteries were loaded onto the sled and an attempt was made to pull it, but the batteries were so heavy it would not move. Since some of the gradients in the ascent they were to make were as steep as 45 degrees, it was immediately apparent that to transport the batteries to the plane was not possible. They did not lose heart, however, because Canessa assured them that it would not be difficult to remove the radio from the pilot's cabin and bring it down to the tail.

Instead of the batteries, then, Canessa and Vizintín piled the sled and filled their knapsacks with warm clothes for the other boys and thirty cartons of cigarettes, while Parrado went back to the galley and wrote above the sink in nail polish, "Go up. Eighteen people still alive." He copied the message twice on other parts of the tail, using the neat lettering he had learned while labeling boxes of nuts and bolts in his father's business, La Casa del Tornillo. Canessa came into the galley to take out the medicine chest they had found there. It contained many different kinds of drugs, including cortisone, which could be used to relieve the asthma from which Sabella and Zerbino both suffered.

When the two came out they found that Vizintín had stepped on and broken the Samsonite sled. This put Parrado into a fury. He swore at Vizintín and cursed him for his clumsiness, but Canessa was able to repair the damage and at last they set out, plodding in their snowshoes over the soft, steep snow toward the plane.

10

The spirits of the boys they had left behind had risen in their absence. There was first of all the immense feeling of relief that at last something was being done about their rescue. They were all quite sure that their expeditionaries would find help. It was also more comfortable in the plane now that they were gone. There was more space to sleep in and less tension without Canessa and Vizintín.

Some of them missed the expeditionaries. Mangino, for example, had lost the protection of Canessa. On the other hand, he now had less need of it because he had grown to be less spoiled. He was a little more stoical about his broken leg and therefore slightly easier to

1

2

3

4

Top: The Old Christians' First Fifteen. Abal and Canessa are seated at the far right; Perez, Vizintin, Harley, and Parrado stand third, fourth, sixth, and seventh from the left.
Left: Nando Parrado
Upper right: Panchito Abal
Lower right: Arturo Nogueira

5

6

Susana Parrado

Marcelo Pérez

Liliana Methol with her children

7

8

A group of the Old Christians and their friends in the streets of Menaoza. From left to right: Valeta, Martínez-Lamas, Mangino, Platero, Zerbino, Inciarte, Turcatti, Magri, and Menéndez

The passengers waiting to board the Fairchild at Mendoza airport

9

10 Inside the Fairchild: Nicolich with his arm around Harley

The track made by the fuselage

11

12

Fito Strauch

13

Eduardo Strauch

14

Daniel Fernández

15

Gustavo Zerbino

16

Roberto Canessa

17

Nando Parrado

18

Antonio Vizintín

Carlitos Páez

19

20

Pancho Delgado

21

Coche Inciarte

22

Pedro Algorta

23

Javier Methol

24

Moncho Sabella

25

Bobby Francois

26

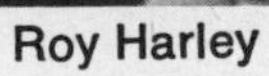

Roy Harley

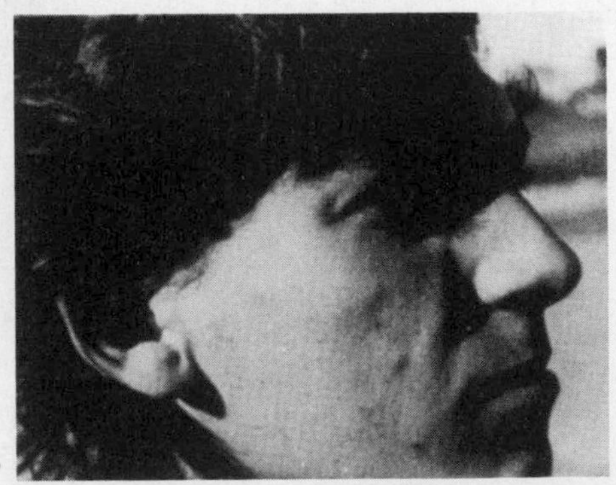

27

Alvaro Mangino

A group of the survivors outside the Fairchild

28

On the mountain

29

Sabella inside the fuselage

Parrado inside the tail section

31

32

Canessa and Vizintín at the tail, Harley with his back to the camera

Sewing the sleeping bag 33

The cross seen on the Santa Elena mountain

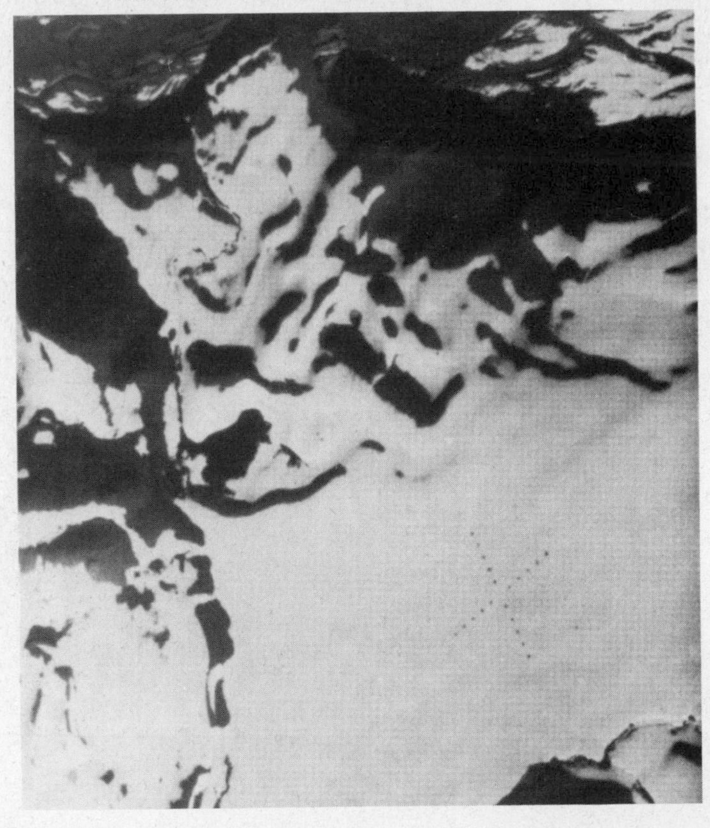

34

35

Canessa and Parrado setting off for Los Maitenes

37

Francois getting out of the helicopter and Sabella inside

Parrado's note to Catalan

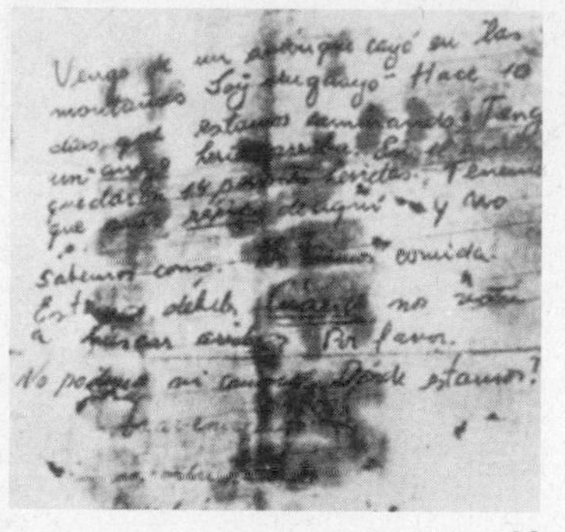

Vengo de un avión que cayó en las
montañas Soy uruguayo- Hace 10
días que estamos caminando. Tengo
un amigo herido arriba. En el avión
quedaron 14 personas heridas. Tenemos
que salir rápido de aquí y no
sabemos como. No tenemos comida.
Estamos débiles. Cuándo nos van
a buscar arriba. Por favor.
No podemos ni caminar. Dónde estamos?

36

Roy Harley reunited with his mother

38

39

The last day on the mountain. From left to right: seated, Methol, Harley, Zerbino, Francois, Sabella, an Andinist, Fito Strauch, and Delgado; standing, Vizintín and two Andinists

The last night in the Fairchild: Fito Strauch, Zerbino, Francois, Harley, Delgado, and Sabella

40

41 Christmas at the Sheraton San Cristóbal

The grave on the mountain

42

sleep with. Methol, his partner—who had once told him that if he was his father he would beat him, so irritated was he by Mangino—now became the confidant of his remorse. "I used to be so spoiled," Mangino said to him. "Being up here makes one realize how dreadful one was before. I used to kick my brother if he annoyed me, or throw away the soup if it wasn't just right. If only I could have that soup now. . . ."

They all felt that they had been through a purifying experience. Delgado, Turcatti, Zerbino, and Fito Strauch once discussed how they were gong through a kind of purgatory. They thought of Christ's forty days in the desert, and since it was now forty days since the plane had crashed they felt sure that their ordeal was about to end; as if to demonstrate that their suffering had indeed made better men of them, they tried harder than ever not to quarrel with one another and to be kind to all.

Certainly their quarrels were never serious when compared to the strong bond of their common purpose. Especially when they prayed together at night they felt an almost mystical solidarity, not only among themselves but with God. They had called to Him in their need and now felt Him close at hand. Some had even come to see the avalanche as a miracle which had provided them with more food.

This union was not just with God but with the friends who had died and whose bodies they were eating to survive. Those souls had been called to heaven because their work on earth had been done, but all who were now living would quite happily have exchanged roles. Nicolich, before the avalanche, and Algorta, while suffocating beneath the snow, had both been prepared to die and bequeath their bodies to their friends. It was also, as Turcatti said to the three others in a conversation about Christ's ordeal in the desert, that their condition on the mountain was so terrible that any other would be better—even death.

Turcatti's spirit continued to decline. He was still bitterly disappointed that he had not been allowed to go on the expedition, and he turned his anger not on the others but on himself. He despised his own weakness and seemed to abandon his own body as a punishment for letting him down. His ration, now that he was no longer an expeditionary, was no larger than that of anyone else, yet even so he would not finish it. He had always found the raw meat repulsive and had only eaten it to build up his strength for the expedition. Without this need, all his repugnance returned—and what he had been willing to do for the sake of all, he was unwilling to do for himself. He therefore put the meat aside and, when the Strauches forced him to eat it, hid it away around him.

As a result, of course, he grew weak and was less able to resist the poison in his leg. Canessa's operation had drained the pus, but the infections grew worse and he took this as an excuse to do less and less for the group or for himself. He would only melt snow for himself and would ask others to do things such as pass a blanket which he was quite capable of doing. The debility of his mind raced ahead of his body's weakness. He once asked Fito to help him stand up. Fito refused, telling him that he could easily do it himself, and sure enough, a few moments later, Turcatti lifted himself from the ground and hobbled into the plane.

It was not that he was angry with the group—he was angry with himself—but he took it out on them as if to say, You're quite right, I'm weak and useless, but just wait and see how weak and useless I can be.

Rafael Echavarren was the reverse. His spirit remained strong but the afflictions of his body slowly showed themselves to be stronger still. Hs wounded leg was now black and yellow from gangrene and pus, and because he could no longer get out of the plane he breathed only the dank air inside, which affected his lungs and gave him difficulty in breathing.

It was cold up on the hammock, and Fernández tried bringing him down onto the floor for the night, but the pain this caused was too great and Echavarren preferred to go back to the hammock. Then one night he became delirious. "Who wants to come with me to the store," he said, "to get some bread and Coca-Cola?" Then he shouted, "Papa, Papa, come in! We're in here. . . ."

Páez went up to him and said, "You can say what you like later, but now you're going to pray with me. 'Hail Mary, full of grace, the Lord is with thee . . .' "

Echavarren's open, staring eyes turned toward Páez, and slowly his lips began to repeat the words of the prayer. For that short moment—the time it took them to say a Hail Mary and an Our Father—he was lucid. Then Páez went to sleep opposite Inciarte, and Echavarren returned to his incoherent raving.

"Who'll come with me to the store?"

"Not me, thanks," they shouted back, or "Let's wait until tomorrow." They were quite hardened to the horror of it all. Soon, however, his delirium ceased; all they could hear was the rasping sound of his labored breathing. Later it quickened and then stopped. Zerbino and Páez leaped up and pushed against his chest to try and start it again. Páez continued this artificial respiration for half an hour, but it was clear to the rest of them after a few minutes that Rafael Echavarren was dead.

Echavarren's death inevitably depressed them; it reminded each one that he too might die. Fito lay with blood seeping from his hemorrhoids, feeling more anxious, frightened, and lonely than ever before, yet just because he sensed his own death to be near, he felt close to God and prayed to Him for his own soul and for the well-being of all the others.

What made his spirits rise again the next morning—as it did for all the fourteen who were left alive in the fuselage—was the thought that at that very moment the expeditionaries might have reached help; that before the next night was upon them he would hear and then see a fleet of helicopters coming to rescue them. All he heard, however, toward evening was the shout of the first of his companions to spot the three figures of the returning expeditionaries, and as Fito himself watched them, plodding and stumbling up the hill, all his pious resignation turned to rage. God had raised their hopes only to dash them; their return was proof that they were trapped. He wanted to scream like a madman and run off into the snow, yet he stood and watched with the others, their faces flaccid with expressions of bitter disappointment and deep despair.

Canessa came first, followed by Parrado and Vizintín. When he came within earshot they could hear his piercing voice shouting, "Hey, boys, we've found the tail . . . all the suitcases . . . clothes . . . and cigarettes," and when he reached the plane they clustered around him and heard what had happened. "Wc wouldn't have made it that way," said Canessa. "The valley doesn't turn, it goes east. But we've found the tail and the batteries. All we have to do is get the radio and take it down there."

In the face of this forceful optimism their spirits lifted. They wept and embraced and then clustered around the sled to pick trousers, sweaters, and socks, while Pancho Delgado took charge of the cigarettes.

Six

1

On the same day as the first expedition left the Fairchild, Madelón Rodríguez and Estela Pérez flew back to Chile. With them went Ricardo Echavarren, the father of Rafael; Juan Manuel Pérez, the brother of Marcelo; and Raul Rodríguez Escalada, the most experienced pilot of Pluna, the Uruguayan National Airline, and a cousin of Madelón. The two Strauch brothers had also intended to be part of the expedition, but both suffered from high blood pressure and were advised to remain behind.

Madelón had been galled by Surraco's skepticism about Gerard Croiset, Jr., and was determined to prove him wrong. Both she and Estela Pérez were as confident now as they had been thirty-six days before, when the plane had first disappeared, that their sons were still alive. The men who were with them were less sanguine about finding survivors, but they thought it important to establish just what had happened to the Fairchild.

On November 18 they reached Talca and began to explore around the Descabezado Grande. First they rented a plane and made several flights around the area.

From the air they saw, by the Laguna del Alto, a piece of country which matched almost exactly the drawing which Croiset had sent them. They flew back to Talca to hire guides and horses and then set off on horseback to return and explore it more closely. The horses were specially trained to pick their way along the narrow mountain paths, with precipitous falls beside them, but the Uruguayans had to bandage their eyes for fear of vertigo. As they approached their goal they found another clue which fitted into Croiset's vision, a sign saying "Danger." There too was the lake and the mountain without a top. It seemed as if at last they were near to finding the Fairchild, yet they felt no exhilaration, for though the valley had some vegetation on which human beings could survive, it was only a day's walk from Talca.

They found nothing and returned to Talca, where a Dutch mountain climber told them, "You may find the plane, but it'll be in the middle of a flock of vultures." It was the sentiment of most people they spoke to.

César Charlone, the Uruguayan chargé d'affaires in Santiago, also seemed to have little faith in their venture. When the group returned to Santiago on November 25, he told them that he was too embarrassed to ask the Chilean government to exempt them from the obligation to change ten U.S. dollars a day into Chilean currency at the unfavorable official rate. Madelón Rodríguez and Estela Pérez were furious. Before they had left Montevideo they had specifically gone to the Uruguayan Foreign Ministry to make sure that Charlone would be told to seek the exemption.

The sum outstanding was five hundred and fifty dollars, which would not have bankrupted the families involved, but the anger of these two spirited women was aroused. Estela Pérez, the daughter of a proud Blanco family (she was the cousin of Blanco leader Wilson Ferreira), did not intend to allow herself to be frustrated by a Colorado. With Madelón she went straight to the Chilean Foreign Ministry, where they were received by the foreign minister himself, Clodomiro Almeyda, who listened to them and then immediately wrote a letter which exempted them from the exchange.

With this exemption arranged, the five were able to return to Montevideo, and they did so on November 25. Páez Vilaró had gone to Brazil, and for almost the first time since the plane had disappeared there was no one searching in Chile. But before they left, they had scattered many of the leaflets offering a reward of 300,000 escudos over the mountains around Talca. They hoped for results from that; otherwise, their energies could only be channeled into prayer.

There were few now who had any faith in young Croiset, and it was around this time that Ponce de León spoke to him for almost the last time. He had said quite recently that when he saw the plane now it was empty, and the mothers had taken this to mean that their sons had set out in search of help, but Rafael began to suspect that it might mean they were all dead. When he had pressed Croiset before, Croiset always said he had lost contact. This time, as he spoke to Croiset, Rafael reminded him of what he had said—that the plane was now empty—and told him of the interpretation the mothers had put on this. "My own feelings," he went on, "are different. I take it to mean that they're dead."

"You can't be sure," said Croiset, over the telephone from the Netherlands.

"Listen," said Rafael. "There's no one here who's related to any of the boys, so tell me frankly what you think has happened to them."

There was a pause. Only the whistling and crackling came from the wire. Then Croiset said, "I think, personally . . . now . . . that they are dead."

This was the end of their contact with Gerard Croiset, Jr., yet those who had at last lost faith in his clairvoyance had not lost faith in the survival of their sons. They turned increasingly to their God and their church. Rosina and Sarah Strauch continued their fervent supplications to the Virgin of Garabandal, and every afternoon, in her house in Carrasco, Madelón would kneel with her mother and her two daughters to pray the rosary. Often they would be joined by Susana Sartori, Rosina Machitelli, and Inés Clerc, three of the *novias* who believed in the return of their future husbands.

Yet parallel to their invocation of the supernatural to help them was their search for a natural explanation of what might have happened to the plane, and it was Roberto's mother, Mecha Canessa, who came up with the idea that the plane might have been hijacked by the Tupamaros and now lay concealed in some secret airstrip in southern Chile, waiting for the right moment for a ransom demand. In the atmosphere of political uncertainty which existed in both Uruguay and Chile at that time, the idea seemed plausible, and inquiries were made into the political background of the pilots. It emerged that they were both men of the Right. Since the idea that any of the boys might have hijacked the plane was discounted, they were left with Señora Mariani. Much was made of the fact that she had been on her way to Chile for the marriage of her daughter to a political exile, but then the absurdity of the idea that this fat middle-aged lady should have commandeered a plane grew to be greater, even, than the need for a new theory.

On December 1 a report appeared in a Montevidean newspaper that the Uruguayan Air Force was about to send a plane to Chile to search for the Fairchild around the Tinguiririca volcano. The story was not officially confirmed. The mothers in Montevideo, encouraged by the story, turned to their husbands and asked why a Uruguayan plane could not be sent at once to do now what the SAR planned to do in two months' time?

On December 5 Zerbino, Canessa, and Surraco, together with Fernández, Echavarren, Nicolich, Eduardo Strauch, and Rodríguez Escalada met with the commander in chief of the Uruguayan Air Force, Brigadier Pérez Caldas. They showed him a full report of what they had done themselves in Chile in the area around Talca but told the brigadier that they no longer had faith in Croiset's visions. The search, they said, must be concentrated in the Tinguiririca area, and as private individuals they had not the means to conduct it.

Pérez Caldas sent for a subordinate who had that day returned from Santiago, where he had been working with the SAR on a technical investigation of the accident. He delivered a report confirming that nothing could be done before February. That winter had seen the heaviest falls of snow in the Andes for the past thirty years. The plane would be completely buried and there was no possibility of survival.

Pérez Caldas turned to the eight men who faced him, expecting them to accept this assessment of the situation, but though in their hearts they all agreed that a search would be fruitless, they insisted that it must take place. They explained the state of mind of the mothers and the *novias,* and the resolve of Pérez Caldas began to weaken. Finally he rose to his feet. “Gentlemen,” he said, “you have made your request and I have made my decision. The Uruguayan Air Force will arrange for a plane to be at your disposal.”

The final search was on its way.

2

A new spirit of optimism was generated among the parents by the news that a specially equipped C-47 of the Uruguayan Air Force was to be sent to search for the Fairchild, and there were many volunteers from among the fathers to join the expedition. Páez Vilaró was still in Brazil, but although the Air Force would allow only five passengers to fly with the crew, it was clear that he would want to be one of them. Ramón Sabella was to have been another, but he was advised against going by his doctor. The four others who were chosen to accompany Páez Vilaró were Rodríguez Escalada and the fathers of Roberto Canessa, Roy Harley, and Gustavo Nicolich. But it was not just these men who were involved in the operation; the families of Methol, Maquirriain, Abal, Parrado, Valeta, and many others contributed money and advice, while Rafael Ponce de León kept in touch with the radio hams in Chile.

On December 8 a group of parents, including those who were to go on the expedition, went to Base No. 1 of the Air Force to confer with the pilot of the C-47, Major Ruben Terra, and draw up plans for the search. On the same day Páez Vilaró returned from Brazil and confirmed that much as he mistrusted the planes of the Uruguayan Air Force, he would fly with the expedition.

They continued throughout the next day to make arrangements for their departure, and at midday on December 10 there was a final meeting of all the parents, relatives, and *novias* with their expeditionaries in the spacious Moorish bungalow of the Nicolichs. Two expert Uruguayan pilots were invited to this meeting, and all the material that had been assembled by the Chilean SAR, the Uruguayan Air Force, and the parents themselves were laid out on the table for all to see. Dr. Surraco produced maps and explained why he and others were now convinced that the plane must have crashed between the Tinguiririca and Sosneado mountains. No one disputed their judgment. The mountains around Talca and Vilches had been forgotten; reason had triumphed over parapsychology.

The meeting went on until evening. After it had broken up the two Strauch couples went to the Harleys' house, where they talked long into the night about the expedition which they hoped so much would find their sons. At the end of it Roy Harley turned to Rosina Strauch and said, "Listen. I'm going to chew up the Andes. I'm going to search them foot by foot until I find the boys. But I'm going to ask you to do something too.

If we fail this time, you must accept that there's no more hope. When we come back from the expedition, we must give up all false expectation."

At six o'clock the next morning, December 11, the C-47 took off for Santiago. Aboard were the pilot, Major Ruben Terra; a crew of four; and Páez Vilaró, Canessa, Harley, Nicolich, and Rodríguez Escalada. Even at that early hour, many parents had come to the airport to see them off.

The plane was a military transport. There were no comfortable seats inside, and the five middle-aged men had to sit on benches along the side. It was noisy too, but they were all quite content because for the first time since the Fairchild had disappeared they had at their disposal the means to search among the highest peaks of the Andes. According to the Uruguayan press, the C-47 had been specially equipped for this expedition; at any rate, it had the oxygen and pressurization required for high-altitude flying. As they flew over the estuary of the River Plate, Páez Vilaró picked up a newspaper that he found in the plane and read an article on the expedition. He looked up, every now and then, to see if he could see the "special equipment of exceptional precision" that the newspaper talked about, and when he came to a passage which talked of the skill and magnificence of the Uruguayan Air Force he became a little irritated. This was, after all, almost the first thing they had done since the disappearance of the Fairchild to look for five of their own men.

He wondered what the pilot thought of the article and went forward into his cabin. He could see, as he entered, that they had reached the Argentinian shore of the estuary and were about to fly over Buenos Aires.

"Have you seen this?" he shouted to Ruben Terra, holding up the newspaper with the article he had just read.

"Yes," said the major.

"What do you think of it?"

"Very fair."

Páez Vilaró gave a slight shrug of his huge shoulders.

The pilot seemed to notice, for he went on, "For instance, once one of these engines died and I was able to land the plane on one—"

At the very moment he spoke the plane gave a sudden lurch and began to vibrate. Páez Vilaró looked back at the wing and saw what the major had just described; the propeller of the starboard engine was coming to a stop.

He turned back to the pilot. "Well, it's happened again," he said.

They made an emergency landing at the military airport of El Palomar, where Ruben Terra cabled to Montevideo for a new engine, but his passengers were not willing to wait until it came and was fitted to the C-47. They hired a light aircraft to fly them to Ezeiza airport and there caught a scheduled LAN Chile flight to Santiago. They arrived around seven in the evening and drove straight to the headquarters of the SAR at Los Cerrillos.

"What?" said Commander Massa when he saw Páez Vilaró and his companions. "You here again? This certainly isn't the right moment to start looking. We told you we'd let you know when it came."

The commander had good reason to be confused. It was not uncommon for an aircraft to disappear in the

Andes, but it was most unusual for anyone to search for it nearly two months later. Even when a DC-3 of the United States Air Force had crashed in 1968 they had not gone on looking for so long, yet here was a group of Uruguayan civilians who moved into his office and refused to be put off.

"Tell me, commander," said Roy Harley. "What are the chances that the plane became trapped in the mountains and made an emergency landing in the snow?"

"With a bit of luck," said Massa, "two in a thousand."

"One is enough for us," said Harley.

The Uruguayans now started to bargain for the use of the SAR's helicopters, but Massa, though he listened politely to what they proposed, could not accept their suggestion. "You don't understand," he said. "It's extremely dangerous flying helicopters in the cordillera. I can't risk the lives of my pilots unless there's some concrete evidence that the wreck is in a specific place. If you have any evidence, give it to me and I'll act upon it, but until then . . . I'm sorry."

The five men returned to their hotel with nothing but this halfhearted assurance from Massa, but they were not discouraged. Though tired from their hectic journey, they immediately sat down to plan. They would divide themselves into groups. One would explore the area of the Palomo and Tinguiririca mountains by land, another would cover the same area from the air as soon as the C-47 reached Santiago, while a third would try to find the copper miner, Camilo Figueroa, who claimed to have seen the plane fall from the sky. These were to be the three prongs of their assault upon the Andes. They named it Operation Christmas.

Seven

1

November 23 was Bobby Francois's twenty-first birthday. He received as a present from his sixteen companions an extra pack of cigarettes. Meanwhile Canessa and Parrado set about the task of removing the radio from the panel of instruments which remained half buried in the pilot's chest.

The earphones and the microphone were connected to a black metal box about the size of a portable typewriter which came out quite easily with the removal of a few screws. They realized, however, that it had no dial and so could only be part of the VHF radio; it also had sixty-seven wires coming out of the back, which they thought must connect it to the missing half. The plane was so full of instruments that it was not easy to decide what might be part of the radio and what might not, but eventually, behind a plastic panel in the wall of the luggage compartment, they found the transmitter. This was much more difficult to get at and separate from all the other instruments—especially as there was no light to work by. Their only tools were a screwdriver, a knife, and a pair of pliers, and with these, after several days of effort, they eventually extracted it.

Their hunch that this was the part which matched the box they had taken from the instrument panel was confirmed by a cable which ran out the back with sixty-seven different wires. The difficulty which faced them was their ignorance of which wire on one piece of equipment matched which wire on the other. With sixty-seven wires on each, there were many million permutations. Then they discovered that the wires had faint markings on them which enabled them to make the correct connections.

Canessa was the most enthusiastic about the radio. He thought it insane for any of them to risk their lives by setting off over the mountains if there was any chance that they could make contact with the outside world. The majority agreed with him, though many were more or less skeptical about the outcome. Pedro Algorta did not think they would ever make it work, but he said nothing which might make the optimists despondent. Roy Harley himself, who was supposed to be their radio expert, was the most doubtful of all. He knew best the limits of his own expertness upon which they based their hopes—some odd afternoons fiddling around with the stereo set of a friend—and insisted repeatedly in a whining voice that this in no way qualified him to dismantle and then reassemble a VHF radio.

The other boys discounted his diffidence as the by-product of his physical and mental debility. His large face was fixed in a permanent expression of misery and despair, and his body, which had once been solid and strong, had shrunk to the wizened dimensions of an Indian fakir. The expeditionaries and the cousins therefore asked him to train for the journey to the tail by walking around the plane, but he was too weak. (They did not consider it appropriate to increase his ration of flesh.) The more insistent they became, the more Roy resisted the idea. He wept and pleaded and told them over and over again that he knew no more about radios than anyone else. Their authority, however, was difficult to resist, and he was under another kind of pressure from another source. "You must go," his friend Francois said to him, "because the radio may be our only chance. If we have to walk out of here—people like Coche, Moncho, Alvaro, you, and me—we just won't make it."

With enormous reluctance Roy gave in to this argument and agreed to go. Their departure, however, was not imminent because several of them were still struggling with the shark's-fin antenna, riveted on the roof of the plane above the pilots' cabin. They had to remove the rivets with only a screwdriver, and the task was made more difficult by twists in the metal caused by the fall of the plane.

Even when it was removed and lay on the snow beside the different parts of the radio, Canessa spent hour upon hour just staring at it and snapped at anyone who asked him what more there was to do that he was not ready to go. The others became impatient, but they were all wary of Canessa's temper. If he had not been an expeditionary—and the most inventive of the three—they might not have put up with him; as it was, they did not wish to antagonize him. All the same, his procrastination seemed unreasonable, and they began to suspect that he was protracting the experiment with the radio to postpone the moment when he might have to set off in the snow.

At last the three Strauch cousins became exasperated. They told him that he must take the radio and go. Canessa could think of no further excuse for delay, and at eight o'clock the next morning a small column assembled for the descent to the tail. First came Vizintín, loaded as usual like a packhorse; then Harley, with his hands in his pockets, and finally the two figures of Canessa and Parrado, with sticks and knapsacks like two winter sportsmen.

They set off down the mountain, and the thirteen they left behind were delighted to see them go. Not only were they spared the irritable and bullying presence of Canessa and Vizintín but also, with the four absent, they could sleep much more comfortably. Above all they could dream again that rescue was at hand.

They were in no position, however, to sit back and wait for their dreams to come true. For the first time since they had taken their decision to eat the flesh of the dead, they were running short of supplies. The problem was not that sufficient bodies did not exist but that they could not find them; those who had died in the accident and had been left outside the plane were now, as a result of the avalanche, buried deep beneath the snow. One or two still remained of those who had died in the avalanche, but they knew that quite soon they would have to find the earlier victims. It was also a consideration that those who had died in the accident would be fatter and their livers better stocked with the vitamins they all needed to survive.

They therefore set about searching for bodies. Carlitos Páez and Pedro Algorta were in charge of this operation, but all the other boys joined in. Their method was to dig a shaft down into the snow on the spot where they remembered a body had lain, but these holes would often go deep without anything coming to light. On other occasions they would be more successful, but often with frustrating consequences. It was thought, for instance, that a body lay somewhere around the entrance to the plane, and Algorta spent many days methodically digging a hole there, with steps going down into it. It was difficult work because the snow was hard and Pedro, like all the others, had grown increasingly weak, so it was something like finding gold when the piece of aluminum which acted as a shovel uncovered the fabric of what seemed to be a shirt. Pedro dug faster around the legs and feet of the body but suddenly saw, as he uncovered them, that the toenails were painted with red varnish. Instead of a boy's body he had found Liliana Methol, and in deference to Javier's feelings they had agreed not to eat her.

Another method of sinking an exploratory shaft was for all the boys to urinate onto a single spot. It was an effective method, if only they could contain themselves each morning for long enough to reach the appointed place. Alas, many of them awoke with bladders so taut that they were forced to relieve themselves as soon as they left the plane. Algorta often slept with his three pairs of trousers unbuttoned, and even then he sometimes did not make it out of the plane. It was a pity, because it was easier to piss into a hole than to dig one.

Many of the boys felt themselves too weak to do any labor of any kind. Some had learned to live with their uselessness, but others did not admit to themselves that they made no contribution to the welfare of the group.

Carlitos once rebuked Sabella for not doing any work, whereupon the enfeebled Moncho fell to digging a hole with such hysterical frenzy that those who watched him feared for his life, but his exertions only led him to collapse with exhaustion. Here indeed was a case where the spirit was willing but the flesh weak. Moncho would have loved to be counted among the heroic cousins and expeditionaries, but his body betrayed him; he had no choice but to be one of the spectators.

At the same time as the boys dug into the snow in search of the buried bodies, the corpses that they had preserved nearer the surface began to suffer from the stronger sun which melted the thin layer of snow which covered them. The thaw had truly set in—the level of the snow had fallen far below the roof of the Fairchild—and the sun in the middle of the day became so hot that any meat left exposed to it would quickly rot. Added, then, to the labors of digging, cutting, and snow melting was that of covering the bodies with snow and then shielding them from the sun with sheets of cardboard and plastic.

As the supplies grew short, an order went out from the cousins that there was to be no more pilfering. This edict was no more effective than most others which seek to upset an established practice. They therefore sought to make what food they had last longer by eating parts of the human body which previously they had left aside. The hands and feet, for example, had flesh beneath the skin which could be scraped off the bone. They tried, too, to eat the tongue off one corpse but could not swallow it, and one of them once ate the testicles.

On the other hand they all took to the marrow.

When the last shred of meat had been scraped off a bone it would be cracked open with the ax and the marrow extracted with a piece of wire or a knife and shared. They also ate the blood clots which they found around the hearts of almost all the bodies. Their texture and taste were different from that of the flesh and fat, and by now they were sick to death of this staple diet. It was not just that their senses clamored for different tastes; their bodies too cried out for those minerals of which they had for so long been deprived—above all, for salt. And it was in obedience to these cravings that the less fastidious among the survivors began to eat those parts of the body which had started to rot. This had happened to the entrails of even those bodies which were covered with snow, and there were also the remains of previous carcasses scattered around the plane which were unprotected from the sun. Later everyone did the same.

What they would do was to take the small intestine, squeeze out its contents onto the snow, cut it into small pieces, and eat it. The taste was strong and salty. One of them tried wrapping it around a bone and roasting it in the fire. Rotten flesh, which they tried later, tasted like cheese.

The last discovery in their search for new tastes and new sources of food were the brains of the bodies which they had hitherto discarded. Canessa had told them that, while they might not be of particular nutritional value, they contained glucose which would give them energy; he had been the first to take a head, cut the skin across the forehead, pull back the scalp, and crack open the skull with the ax. The brains were then either divided up and eaten while still frozen or used to make the sauce for a stew; the liver, intestine, muscle, fat, heart, and kidneys, either cooked or uncooked, were cut up into little pieces and mixed with the brains. In this way the food tasted better and was easier to eat. The only difficulty was the shortage of bowls suitable to hold it, for before this the meat had been served on plates, trays, or pieces of aluminum foil. For the stew Inciarte used a shaving bowl, while others used the top halves of skulls. Four bowls made from skulls were used in this way—and some spoons were made from bones.

The brains were inedible when putrid, so all the heads which remained from the corpses they had consumed were gathered together and buried in the snow. The snow was also combed for other parts which had previously been thrown away. Scavenging took on an added value—especially for Algorta, who was the chief scavenger among them. When he was not digging holes or helping the cousins cut up the bodies, his bent figure could be seen hobbling around the plane, poking into the snow with an iron stick. He looked so much like a tramp that Carlitos gave him the nickname of Old Vizcacha*—but his dedication was not without its rewards, because he found many pieces of old fat, some with a thin strip of flesh. These he would place on his

* . . . a land-louping scapegrace, hung with rags,
That lived like a leech in the fens and quags,
A gully-raking veteran scamp,
Bad-biled as a mangy boar.

—José Hernández,
The Gaucho Martín Fierro, II, xiv,
translated by Walter Owen
(Oxford: Basil Blackwell, 1935)

part of the roof. If they were waterlogged, they would dry out in the sun, forming a crust which made them more palatable. Or, like the others, he would put them on a piece of metal which caught the sun and in this way warm what he was eating; once, when the sun was exceptionally strong, he actually cooked them.

It was a relief to Algorta that the expeditionaries were no longer there to help themselves to what he had so carefully scavenged and prepared. On the other hand, he did share what he had found with Fito. Alongside his bit of the roof there was Fito's and between the two was an area for food which they shared. It was Algorta who kept this common larder stocked with extra food. He had attached himself to Fito in the same way that Zerbino had become, as Inciarte put it, "the German's page boy." It particularly annoyed Inciarte that Zerbino would give cigarettes to Eduardo even when he still had some of his own, but Zerbino remembered the days after the first expedition, when Eduardo had let him sleep with his swollen feet on his shoulders.

As the division between the two groups of workers and work-shy, provident and improvident, grew wider, Coche Inciarte's role became more important. By performance and inclination he was firmly in the camp of the parasites; on the other hand he was an old friend of Fito Strauch and Daniel Fernández. He also had the kind of pure and witty character that it was impossible to dislike. Whether he was coaxing Carlitos to cook on a windy day or shooting a pint of pus out of his dreadfully infected leg, he would always smile himself and make others smile at what he was doing. His condition, like Numa Turcatti's, was increasingly serious because both were reluctant to eat raw meat. Coche even became delirious at times and told the boys in all seriousness that there was a little door in the side of the plane where he slept which led out into a green valley. Yet when he announced one morning, as Rafael Echavarren had done, that he was going to die that day, no one took him seriously. Next day, when he awoke again, they all laughed at him and said, "Well, Coche, what's it like to be dead?"

Cigarettes were, as always, the chief source of tension. Those like the cousins, who had sufficient control of themselves to space their smoking and make their ration last, would find toward the end of the second day that each puff was watched by a dozen pairs of envious eyes. The improvident—and Coche was among the most improvident—would exhaust their own supply the first day and then try and bum cigarettes off those who still had some. Pedro Algorta, who smoked less than the others, would move around with lowered eyes for fear of intercepting one of Coche's importuning glances, yet if he avoided them for long enough, Coche would say to him, "Pedro, when we get back to Montevideo, I'll invite you to eat some *gnocchi* at our uncle's house," at which the hungry Algorta would look up and be caught by the large, pleading, laughing eyes.

Pancho Delgado was also unable to ration himself and would sidle up to Sabella, say, and talk to him about his school days with his brother on the off chance that a cigarette might come out of it. Or he would be sent by Inciarte to persuade Fernández to give them an advance ration. "You see," he would say, "Coche and I are especially nervous people. . . ."

There had been a time when Delgado himself had been in charge of the cigarettes, which was something like putting an alcoholic in charge of a bar. One night the storm outside was so dreadful that snow blew right into the plane. Delgado and Zerbino, who had been sleeping by the door, moved up the cabin to talk and smoke cigarettes with Coche and Carlitos. Most of the boys had stayed awake, smoking and listening to the rumble of avalanches, but in the morning, when they awoke all white from the snow, some doubted that they had smoked quite as many cigarettes as Pancho maintained.

There was another occasion when Fernández and Inciarte quarreled over cigarettes. Fernández, who had charge of one of the three lighters, ignored Coche when he repeatedly asked to use it, because he thought Coche smoked too much. This made Coche furious, and for the rest of that day he refused to speak to Fernández. At night, as usual, they lay next to each other, but every time Fernández's head lolled over onto Coche's shoulder, Coche would shrug it off. Then Fernández said, "Come on, Coche," and Coche gave up his petulant rage. He was too good-natured to sustain it.

Their improvidence with cigarettes reinforced a bond which had already existed between Coche and Pancho. They either bummed alone—Pancho taking some of Numa Turcatti's ration because he felt that smoking was doing him no good, while Coche tried to catch Algorta's eye—or, as we have seen, presented a common front to get an advance from Daniel Fernández. They also talked together about life in Montevideo and weekends in the country with Gaston Costemalle, who had been their mutual friend. Pancho, with his natural eloquence, described the scene of their former happiness so well that Coche would be transported away from the damp, stinking confines of the wrecked plane to the green pastures of his dairy farm. Then, when the story ended, he would suddenly find himself back in the foul reality, which would so depress him that he would sit like a corpse with glazed eyes.

Because of this the Strauches and Daniel Fernández tried to keep Coche away from Delgado. They felt that these escapist conversations would lower his morale to such an extent that he would lose the will to survive. Also, they were coming increasingly to mistrust Delgado. There was an incident when some of the boys were outside the plane and called to those inside to send someone out to fetch their ration of meat. Pancho appeared and, while taking the pieces of meat, asked Fito if he could take a piece for himself.

"Of course," said Fito.

"The best piece?"

"If you like."

Fito and the others had remained on the roof eating their portions of flesh, and after delivering the rest to the others, Pancho had come out to join them. When Fito had eventually gone in, Daniel Fernández, who had cut the meat Pancho brought into smaller pieces, said to him, "Hey, you didn't give us much to eat."

"I cut twelve pieces," said Fito.

"More like eight. I had to cut them all up again."

Fito shrugged his shoulders and said nothing more, for to have expressed what he suspected would have been against his better judgment. It was essential to the group that there should be no real dissension.

Carlitos, on the other hand, felt less compunction. "I wonder where the ghost is, then," he said, staring at Pancho, "who took the four other pieces?"

"What do you mean?" said Pancho. "What are you suggesting? Don't you trust me?"

They might have said more, but Fito and Daniel Fernández told Carlitos to let the matter go.

2

While these developments were taking place in the plane, the three expeditionaries and Roy Harley were in the tail. Their journey down had only taken them one and a half hours, and on the way they had found a suitcase which had belonged to Parrado's mother. They found candy inside and two bottles of Coca-Cola.

They spent the rest of that first day at the tail resting and looking through the suitcases which had appeared from under the melting snow since they were last there. Among other things Parrado found a camera loaded with film and his airline bag with the two bottles of rum and liqueur which his mother had bought in Mendoza and asked him to carry for her. Neither was broken, and they opened one of them but saved the other for the expedition they would have to make if they could not get the radio to work.

Canessa and Harley set about that task next morning. It seemed at first that it would not be difficult, because the sockets in the back of the transmitter were marked BAT and ANT to show where the wires to the batteries and antenna should be fixed. Unfortunately there were other wires whose connections were not so clear. Above all they could not make out which wires were positive and which negative, so often when they made a connection sparks flew into their eyes.

Their hopes of success were raised when Vizintín found an instruction manual for the Fairchild lying in the snow beside the tail. They looked at the index for some reference to the radio and discovered that the whole of chapter thirty-four was devoted to "Communications," but when they came to look for this section they discovered that certain pages had been torn out of the book by the wind and it was just these pages which made up the chapter they needed.

They had no choice, therefore, but to return to trial and error. It was a painstaking business, and while the others worked Parrado and Vizintín would rummage around rifling all the baggage for a second time, or light a fire to cook the meat. Though there were only four of them, they were not exempt from the tensions which existed up in the fuselage. It irritated Roy Harley, for example, that Parrado would not give him the same ration as the others. It seemed clear that since he was on an expedition he should eat the same amount as an expeditionary. Parrado, on the other hand, held that Roy was only an auxiliary; if the radio failed he would not have to walk out through the mountains. Therefore he should eat only what was necessary to survive.

Nor would he let Roy smoke cigarettes. His reasoning was that they only had one lighter with them and they would need it for any final expedition; but it was also true that neither Parrado, Canessa, nor Vizintín smoked themselves but all were intolerably irritated by

Roy's whimpering and wailing. Thus they told him that he could only smoke when they lit a fire. On one occasion, however, when Roy came to light a cigarette at the fire, Parrado, who was cooking, told him to get out of his way and come back when he had finished. But when Roy came back the fire had gone out. He was so angry that he picked up the lighter which Parrado had left on some cardboard and lit a cigarette. When the three expeditionaries saw what he had done they went for him like a pack of zealous school prefects. They cursed him and might have snatched the cigarette out of his mouth had not Canessa thought better of it and stopped them. "Leave him alone," he said to the other two. "Don't forget that Roy may be the one to save the lives of us all by getting this damned radio to work."

It was clear by the third day that they had not brought enough meat to last them the time it would take to set up the radio. Therefore Parrado and Vizintín set off for the plane again, leaving Harley and Canessa at the tail. The ascent, as before, was a thousand times more difficult than the descent had been. After reaching the top of the hillock which lay just east of the plane, Parrado was assaulted by a momentary but profound despair; instead of the fuselage and its thirteen inhabitants, there was nothing but a huge expanse of snow.

He assumed at once that there must have been another avalanche which had completely covered the plane, but he looked up and saw no signs of a fresh fall of snow on the sides of the mountains above him. He walked on and to his immense relief found the plane on the other side of the next hill.

The boys had not expected them and had no meat

prepared. Also, they were all almost too weak to dig for the bodies that would have to be found if the expeditionaries were to restock their larder. Therefore Parrado and Vizintín themselves set about digging. They found a corpse from which the cousins cut meat and stuffed it into rugby socks, and after two nights up in the plane they returned to the tail.

There they found that Harley and Canessa had made all the necessary connections between battery and radio and radio and shark's-fin antenna but still could not pick up any signal on the earphones. They thought that perhaps the antenna was faulty, so they tore out strands of cable from the electrical circuits of the plane and linked them together. One end of this they tied to the tail, the other to a bag filled with stones which they placed on a rock high on the side of the mountain, making an aerial more than sixty feet long. When they connected it to the transistor radio which they had brought with them, they could pick up many radio stations in Chile, Argentina, and Uruguay. When they connected it to the Fairchild's radio, however, nothing came through at all. They therefore reconnected the transistor, found a program which played some cheerful music, and went back to work.

Suddenly there was a shout from Parrado; he had found in one of the suitcases a photograph of a child at a birthday party. It was a little girl, and she was sitting at a table piled with sandwiches, cakes, and crackers. Parrado clutched the photograph and devoured the food with his eyes, but soon the other three, alerted by his cry, came up behind him and joined in the feast. "Just look at that cake," said Canessa, groaning and rubbing his stomach.

"What about the sandwiches?" said Parrado. "I think I'd even rather have the sandwiches."

"The crackers," Vizintín moaned. "Just give me the crackers. . . ."

On the transistor radio that they had attached to their antenna, the four of them heard a news bulletin in which it was announced that the search was to be resumed by a Douglas C-47 of the Uruguayan Air Force. They received the news in different ways. Harley was ecstatic with hope and joy. Canessa too looked relieved. Vizintín showed no particular reaction, while Parrado looked almost disappointed. "Don't get too optimistic," he warned the others. "Just because they're looking again doesn't mean they'll find us."

They decided, all the same, that it would be a good idea to make a large cross in the snow by the tail, and they did so with the suitcases that lay scattered all around. By now they had almost given up hope of the radio, though Canessa still pottered with it and prevaricated about a return to the plane. Parrado and Vizintín, on the other hand, already had their minds on the expedition, for it had been decided up in the plane that if the radio failed the expeditionaries would set off straight up the mountain in obedience to the only thing of which they were sure—that Chile was to the west. Thus Vizintín removed the rest of the material which was wound around the Fairchild's heating system in the dark locker at the base of the tail which had contained the batteries. It was light and yet designed by the most technologically sophisticated industry in the world to contain heat; sewn together into a large sack, it would make an excellent sleeping bag and solve the one outstanding problem which had beset them—how to keep warm at night without the shelter of either tail or plane.

All the time they had been at the tail the snow had been melting around them—all, that is, except the snow immediately beneath the tail, which was shaded from the sun. As a result, the tail had been left as if on a pillar of snow which made it not only more difficult to enter but also perilously unsteady as they moved about inside. On the last night it wobbled so much that Parrado became terrified that it would fall over and shoot down the mountainside. The four lay as still as they could, but there was a wind which made the tail sway and Parrado could not sleep. "Hey," he said at last to the others, "don't you think we'd better sleep outside?"

Vizintín grunted and Canessa said, "Look, Nando. If we're going to die, we're going to die, so let's at least get a good night's sleep."

The tail was in the same place the next morning, but it was clearly no longer safe. It was equally obvious that no more tinkering with the radio would make it work. They therefore made up their minds to return to the plane. Before they left they loaded themselves once again with cigarettes, and Harley—as an expression of all the misery and frustration he had felt in those eight days—kicked to pieces the different components of the radio they had so painstakingly put together.

He was wrong to waste his energy. The 45-degree climb back to the plane was almost a mile. At first it was not so difficult, because the surface of the snow was hard. Later, when it became mushy and they either sank up to their thighs or strapped on the heavy and cumbersome cushions that they used as snowshoes, it needed an almost super-human effort which poor Roy was not in a condition to provide. Though they rested every thirty paces, he soon lagged behind, but Parrado stayed with him—cajoling, cursing, begging him to come on. Roy would try and then fall back into the snow with despair and exhaustion. His voice whined higher than ever before; tears flowed more freely. He begged to be left there to die, but Parrado would not leave him. He swore at him and insulted him to put fire into his blood. He cursed him as he had never cursed anyone before.

The insults were extreme but they worked. They drove Roy on until it came to a point where he no longer responded to either oath or insult. Then Parrado returned to him and spoke quietly, saying, "Listen. It isn't much farther now. Don't you think it's worth making one last little effort for the sake of seeing your mother and father again?"

Then he took his hand and helped Roy to his feet. Once again he staggered up the mountain, resting on Parrado's arm, and when they came to a slope of snow so steep that no effort of will could drive Roy to surmount it, Parrado gripped him with the enormous strength he still seemed to possess and brought him up toward the Fairchild.

They reached the plane between half past six and seven in the evening. There was a cold wind blowing, with a slight flurry of snow. The thirteen had already gone inside, and they gave the expeditionaries a depressed reception.

Canessa, however, was less struck by their unfriendliness than by the desolation of the spectacle they presented. After eight days away he saw with some objectivity just how thin and haggard the bearded faces of his friends had become. He had seen too, with a fresh eye, the horror of the filthy snow strewn with gutted carcasses and split skulls, and he thought to himself that before they were rescued they must do something to tidy it up.

3

Toward the end of the first week in December, after fifty-six days on the mountain, two condors appeared in the sky and circled above the seventeen survivors. These two enormous birds of prey with bald necks and head, a collar of white down, and a wingspan of nine feet were the first sign of any life but their own that they had seen for eight weeks. The survivors were immediately afraid that they would descend and carry off the carrion. They would have shot at them with the revolver but were afraid that the sudden noise might cause another avalanche.

At times the condors would leave them but then reappear the next morning. They watched the movements of the human beings but never swooped on them, and after some days they disappeared altogether. They were followed, however, by other signs of life. A bee once flew into the fuselage and then out again; later still, one or two flies and finally a butterfly were seen around the plane.

It was now warm during the day; indeed, at midday it was so hot that their skin became burned and their lips were cracked and bleeding. Some of them tried to make a tent to shelter them from the sun with the poles from the hammocks and a bale of cloth that Liliana Methol had bought in Mendoza to make a dress for her daughter. They also thought it would make a useful signal to any plane that flew overhead, for this possibility was uppermost in their minds. When Roy and the expeditionaries had returned from the tail, they had told the thirteen who had stayed in the plane that they had heard on their transistor that the search had been restarted.

The boys were determined that this should not tempt the expeditionaries to abandon the idea of further expedition. They had had no high hopes for the radio and were not thrown into despair when Harley, Canessa, Parrado, and Vizintín returned, but they were impatient that the last three should leave again almost immediately. It soon became evident, however, that while the news of the C-47 in no way affected Parrado's determination to leave, it produced in Canessa a certain reluctance to risk his life on the mountainside. "It would be absurd for us to leave now," he said, "with this specially equipped plane on its way to find us. We should give them at least ten days and then, perhaps, set out. It's crazy to risk our lives if it isn't necessary."

The others were thrown into a fury by this procrastination. They had not pampered Canessa and suffered his intolerable temper for so long only to be told by him that he was not going. Nor were they so optimistic that the C-47 would find them, for they heard on the radio first that it had been forced to land in Buenos Aires, then that it had had to have its engines overhauled in Los Cerrillos. There was also the shortage of food that was upon them, for though they knew that corpses were hidden in the snow beneath them, they either could not find them or only found those they had agreed not to eat.

There was another factor too, which was their sense of pride in what they had already achieved. They had survived now for eight weeks in the most extreme and inhuman conditions. They wanted to prove that they could also escape on their own initiative. They all loved to think of the expression on the face of the first shepherd or farmer they found as he was told that the three expeditionaries were survivors from the Uruguayan Fairchild. All of them practiced in their minds the nonchalant tone they would adopt when telephoning their parents in Montevideo.

Fito's impatience was more practical. "Don't you realize," he said to Canessa, "that they aren't looking for survivors? They're looking for dead bodies. And the special equipment they talk about is photographic equipment. They take aerial photographs and then go back, develop them, study them. . . . It'll take weeks for them to find us, even if they do fly directly overhead."

This argument seemed to convince Canessa. Parrado did not need to be persuaded, and Vizintín always went along with what the other two decided. They therefore set to work to prepare the final expedition. The cousins cut flesh off the bodies not just for their daily needs but to set up a store for the journey. The others were set to sewing the insulating material from the tail into a sleeping bag. It was difficult. They ran out of thread and had to use wire from the electric circuits.

Parrado would have helped them do this, but he was not skilled with his hands. He therefore took photographs with the camera they had found in the tail and collected together the clothes and equipment he would need for the expedition. He filled a knapsack made from a pair of jeans with the plane's compass, his mother's rug, four spare pairs of socks, his passport, four hundred U.S. dollars, a bottle of water, a pocket knife, and a woman's lipstick for his broken lips.

Vizintín put his shaving kit into his knapsack, not so much because he intended to shave before reaching civilization but because it had been a present from his father and he did not want to leave it behind. He also packed the plane's charts, a bottle of rum, a bottle of water, dry socks, and the revolver.

Canessa's rucksack was filled with all the medicines he thought they might conceivably need on their journey—adhesive bandages, a tube of dental floss, aspirin, pills against diarrhea, antiseptic cream, caffeine pills, muscle-warming ointment, and a large pill whose function was not known. He added to this a woman's foundation cream to protect his skin, toothpaste, his documents, including his vaccination certificate, Methol's penknife, a spoon, a piece of paper, a length of wire, and an elephant hair as a good-luck charm.

December 8 was the Feast of the Immaculate Conception. To honor the Virgin, and to persuade her to intercede for the success of their final expedition, the boys in the Fairchild decided to pray the full fifteen mysteries of the rosary. Alas, soon after they had finished five their voices grew thinner and fewer, and one by one they dropped off to sleep. They therefore made up the rest the next evening, the ninth, which was also Parrado's twenty-third birthday. It was a mildly melancholy occasion, for they had so often planned the party they would have in Montevideo. To celebrate here on the mountain, the community gave Parrado one of the Havana cigars that had been found in the tail. Parrado smoked it, but he took more pleasure from the warmth it provided than from the aroma.

On December 10, Canessa still insisted that the expedition was not ready to leave. The sleeping bag was not sewn to his satisfaction, nor had he collected together everything he would need. Yet instead of applying himself to what was still to be done, Canessa lay around "conserving his energy" or insisted on treating the boils that Roy Harley had developed on his legs. He also quarreled with the younger boys. He told Francois that at the tail Vizintín had wiped his backside on Bobby's best Lacoste T-shirt, which put Bobby into an unusual rage. He even quarreled with his great friend and admirer Alvaro Mangino, for that morning, while defecating onto a seat cover in the plane (he had diarrhea from eating putrid flesh), he told Mangino to move his leg. Mangino said that it had been cramped all night and so he would not. Canessa shouted at Mangino. Mangino cursed Canessa. Canessa lost his temper and grabbed Mangino by the hair. He was about to hit him but simply threw Alvaro back against the wall of the plane instead.

"Now you're not my friend any more," Mangino said, sobbing.

"I'm sorry," said Canessa, sitting back, his temper once more under control. "It's just that I'm feeling so ill. . . ."

Hc was no onc's fricnd that day. Thc cousins thought he was deliberately procrastinating and were especially angry with him. That night they did not keep his special place as an expeditionary, and he had to sleep by the door. The only one who had any influence on him was Parrado, and his determination was as great as it had ever been. That morning, as they lay in the plane waiting to go out, he suddenly said, "You know, if that plane flies over us, it might not see us. We should make a cross." And without waiting for anyone else to take up his idea, he went out of the plane and surveyed an area of pristine snow where a cross could be best constructed. The other boys followed him, and soon all those who could walk without pain were stamping the ground along pre-ordained lines to make a giant cross in the snow.

In the middle, where the two lines crossed, they put the upturned trash can which Vizintín had brought up on the trial expedition. They also laid out the bright yellow and green jackets of the pilots. Realizing that movement would attract a flier's attention, they drew up a plan whereby they would run in circles as soon as a plane was sighted overhead.

That night Fito Strauch went to Parrado and said that if Canessa would not leave on the expedition then he would.

"No," said Parrado. "Don't worry. I've talked to Muscles. He'll go. He must go. He's much better trained than you are. All we have to do now is finish the sleeping bag."

The next morning the Strauches rose early and set to work on the sleeping bag. They were determined that by that evening there should be no possible excuse for any further delay. But something was to happen that day which would make their threats and admonitions superfluous.

Numa Turcatti had been getting weaker every day. His health, along with that of Roy Harley and Coche Inciarte, caused the greatest concern to the two "doctors," Canessa and Zerbino. Though Numa, who was so pure in spirit, was loved by all on the plane, his closest friend before the accident had been Pancho Delgado, and it was Delgado who took it upon himself to look after him. He brought Numa's ration of food into the plane, made water for him, tried to stop him from smoking cigarettes because Canessa had said they were bad for him, and fed him little smears of toothpaste from a tube which Canessa had brought from the tail.

For all this care Numa continued to decline, and Delgado decided that something further should be done. He made up his mind to get extra food for his patient and, true to his nature, fell upon the method of stealth. He felt, perhaps, that had he asked the cousins outright, they might have refused. There came a day, however, when Canessa had diarrhea and was sitting near where Numa lay inside the plane. Delgado went out to fetch the food and came in with three dishes. Canessa had decided not to eat, because of his diarrhea, but it was a day when they had cooked a stew, and when he saw what Delgado had brought in he asked to taste it. Delgado let him, willingly; Canessa tasted it and then decided that after all he would have his ration.

Canessa went up to Eduardo, who was serving it out, and asked for his ration. Eduardo said, "But I gave your portion to Pancho."

"Well, he never gave it to me."

Eduardo was quick to lose his temper, and he lost it now. He began to vilify Delgado, and as he did so Delgado came out of the plane.

"Are you talking about me?"

"I am. Did you think we wouldn't notice how you sneaked an extra ration?"

Delgado blushed and said, "I don't know how you can think such a thing of me."

"Then why didn't you give Muscles his ration?"

"Do you think I kept it for myself?"

"Yes."

"It was for Numa. You may not realize it, but Numa is growing weaker every day. If he doesn't have extra food he will die."

Eduardo was taken aback. "Then why didn't you ask us?"

"I thought you might refuse."

The cousins let it pass, but they remained suspicious of Delgado. They knew, for instance, that on the days when the meat was raw Numa could hardly be persuaded to eat one ration, let alone two. Nor did it escape their notice that those cigarettes Delgado so conscientiously prevented Numa from smoking, he smoked himself.

Even with an extra ration, Numa's condition did not improve. Instead he grew weaker. As he grew weaker, he grew more listless, and as he grew more listless he bothered less about feeding himself, which in its turn made him weaker still. He also developed a bedsore on his coccyx, and it was when he asked Zerbino to come and look at it that Zerbino realized how drastically thin he had become. Until then his face had been covered by a beard and his whole body by clothes. Removing the clothes to examine the sore, Zerbino could see that there was practically no flesh left between the skin and the spine. Numa had become a skeleton, and Zerbino told the others afterward that he only gave him a few more days to live.

Like Inciarte and Sabella, Numa was intermittently delirious, but on the night of December 10 he slept peacefully. In the morning Delgado went out to sit in the sun. He had been told that Numa might die but his mind would not accept it. Later in the morning, however, Canessa came out and told him that Numa was in a coma. Delgado returned immediately to the cabin and went to the side of his friend. Numa lay there with his eyes open, but he seemed unaware of Delgado's presence. His breathing was slow and labored. Delgado knelt beside him and began to say the rosary. As he prayed, the breathing stopped.

At midday the cushions were laid on the floor of the plane again. It had become a habit, because of the heat of the sun, to take a siesta. It depressed them all to lie impotently like this, but it was better than burning. They sat and talked or dozed off. Then, at about three in the afternoon, they began to file out again. On this particular afternoon, Javier Methol lay at the back of the plane. "Be careful," he said to Coche as he rose and stepped over Numa's body. "Be careful not to step on Numa."

"But Numa's dead," said Parrado.

Javier had not realized what had happened, and now that he understood his spirits dropped completely. He wept as he had wept at the death of Liliana, for he had grown to love the shy and simple Numa Turcatti as though he were his brother or son.

Turcatti's death achieved what argument and exhortation had failed to achieve; it persuaded Canessa that they could wait no longer. Roy Harley, Coche Inciarte, and Moncho Sabella were all weak and incipiently delirious. A day's delay could mean the difference between their death and their survival. It was therefore agreed by all that the final expedition should set off the next day due west to Chile.

That evening, before he went into the plane for the last time, Parrado drew the three Strauch cousins to one side and told them that if they ran short of food they should eat the bodies of his mother and sister. "Of course I'd rather you didn't," he said, "but if it's a matter of survival, then you must."

The cousins said nothing, but the expression on their faces betrayed how moved they were by what Parrado had said to them.

4

At five o'clock the next morning, Canessa, Parrado, and Vizintín prepared to go. First they dressed themselves in the clothes they had picked from the baggage of the forty-five passengers and crew. Next to his skin Parrado wore a Lacoste T-shirt and a pair of long woolen woman's slacks. On top of these he wore three pairs of jeans, and on top of the T-shirt six sweaters. Next he put a woolen balaclava over his head, then the hood and shoulders that he had cut from Susana's fur coat, and finally a jacket. Under his rugby boots he wore four pairs of socks which he covered with plastic supermarket bags to keep out the wet. For his hands he had gloves; for his eyes a pair of sunglasses; and to help him climb he held an aluminum pole which he strapped to his wrist.

Vizintín also had a balaclava. He wore as many sweaters and pairs of jeans but covered them with a raincoat, and on his feet he wore a pair of Spanish boots. As before he also carried the heaviest load, including a third of the meat, packed either in a plastic bag or a rugby sock. With it there were pieces of fat, to provide energy, and liver, to give them vitamins. The whole supply was designed to last the three of them ten days.

Canessa carried the sleeping bag. To cover his body and keep it warm he had looked for woolen clothes, feeling that elemental conditions called for elemental materials. He also liked to think that each garment had something precious about it. One of the sweaters he wore had been given to him by a dear friend of his mother, another by his mother herself, and a third had been knitted for him by his *novia,* Laura Surraco. One of the pairs of trousers he wore had belonged to his closest friend, Daniel Maspons, and his belt had been given to him by Parrado with the words, "This was a present from Panchito, who was my best friend. Now you're my best friend, so you take it." Canessa accepted this gift; he also wore Abal's skiing gloves and the skiing boots which belonged to Javier Methol.

The cousins gave the expeditionaries some breakfast before they sent them on their way. The others watched in silence. No words could express what they felt at this awesome moment; they all knew that this was their last chance of survival. Then Parrado separated once again the pair of tiny red shoes that he had bought in Mendoza for his nephew. He put one in his pocket and hung the other from the hat rack in the plane. "I'll be back to get it," he said. "Don't worry."

"All right," they all said, their spirits raised high by his optimism. "And don't forget to book us rooms in the hotel in Santiago." Then they embraced, and amid cries of *¡Hasta luego!* the three expeditionaries set off up the mountain.

After they had gone about five hundred yards, Pancho Delgado came bobbling out of the plane. "Wait," he shouted, waving a small statue in his hand, "you've forgotten the Virgin of Luján!"

Canessa stopped and turned back. "Don't worry!" he shouted back. "If she wants to stay, let her stay. We'll go with God in our hearts."

They climbed up the valley, but they knew that this course took them slightly northwest and at some moment they would have to turn due west and climb directly up the mountain. The difficulty was that the slopes which encircled them looked uniformly steep and high. Canessa and Parrado began to argue about when, and how soon, they should start to climb. Vizintín, as usual, had no opinion on the subject. Eventually the two agreed. They took a reading on the plane's spherical compass and started to climb due west up the side of the valley. It was very heavy going. Not only were they faced with the steep slope, but the snow had already started to melt and even in their improvised snowshoes they sank up to their knees. The wet snow also made the cushions sodden and therefore exceptionally heavy to drag bowlegged up the mountain. But they persevered, pausing every few yards for a short rest, and by the time they stopped by an outcrop of rocks for lunch at midday they were already very high. Beneath them they could still see the Fairchild, with some of the boys sitting on the seats in the sun watching their progress.

After their meal of meat and fat, they took another short rest and then continued on their way. Their plan was to reach the top before dark, for it would be almost impossible to sleep on the steep slopes of the mountain. As they climbed, their minds were on the view they hoped to have on the other side—a view of small hills and green valleys, perhaps with a shepherd's hut or a farmhouse already in sight.

As they had already found out, however, distances in the snow were deceptive, and by the time the sun went behind the mountain they were still nowhere near the top. Realizing that somehow they would have to sleep on the mountainside, they started to look for a level surface. To their growing dismay, it seemed that there was none. The mountain was almost vertical. Vizintín climbed an outcrop of rock (to avoid going around it in the snow) and got stuck. He very nearly toppled off because of the weight of his knapsack, and only saved himself by untying it and throwing it down onto the snow. The experience unnerved him and he started to moan that he could not go on. He was totally exhausted; to move his legs he had to lift them with his hands.

It was growing dark, and a feeling of panic was coming over them all. They came to another outcrop of rock. Parrado thought that there might be a level surface on the top and started to climb it, while Canessa waited beneath with his knapsack. Suddenly Canessa heard a shout of "Look out!" and a large rock, dislodged by Parrado's rugby boots, came hurtling past him, narrowly missing his head. "For God's sake," Canessa shouted up. "Are you trying to kill me?" Then he started to weep. He felt utterly depressed and in despair.

There was nowhere for them to sleep at the top of the outcrop, but a little farther on they came to an immense boulder, beside which the wind had blown a trench in the snow. The floor of the trench was not horizontal, but the wall of snow would prevent them from slipping down the mountainside; they therefore pitched camp and climbed into the sleeping bag.

It was a perfectly clear night and the temperature had sunk to many degrees below freezing, but the sleeping bag succeeded in keeping them warm. They also ate some more meat and drank a mouthful each of the rum they had brought with them. The view from where they lay was magnificent. There spread before them a huge landscape of snow-covered mountains lit by the pale light of the moon and stars. They felt strange lying there—Canessa in the middle—half possessed by terror and despair, yet half marveling at the magnificence of this icy beauty before them.

At last they slept or slipped fitfully into semiconsciousness. The night was too cold and the ground too hard for the three to sleep well, and the first light of the morning found them all awake. It was still cold and they remained in the sleeping bag, waiting for the sun to appear over the mountains and thaw out their boots, which had frozen solid on the rock where they had left them. While they waited they drank water from the bottle, ate some meat, and took another mouthful each of the rum.

They all watched the changing landscape as it grew lighter, but Canessa's eyes, which were the best of the three, came to concentrate on a line along the valley to the east, far beyond the Fairchild and the tail. Since the whole area was still in shadow it was hard to tell, but it seemed to him that the ground there was not covered by snow and that the line which crossed it might be a road. He said nothing about it to the others, because the idea was absurd; Chile was to the west.

When the sun came up from behind the mountains opposite them they started to climb once again—Parrado first, followed by Canessa and then Vizintín. All three were still tired and their limbs were stiff from the exertion of the day before, but they found a kind of path in the rock which seemed to lead toward the summit.

The mountain had become so steep that Vizintín did not dare look down. He simply followed Canessa at a cautious distance, as Canessa followed Parrado. What frustrated them all was that each summit they saw above them turned out to be a false one, a ridge of snow or an outcrop of rocks. They stopped by one of these rocks to eat in the middle of the day, took a short rest, and then climbed on. By the middle of the afternoon they still had not reached the top of the mountain—and though they felt themselves to be near, they were afraid of making the same mistake as the night before. They therefore looked for and found a similar trench carved by the wind beside the same kind of rock and decided to stop there.

Unlike Vizintín, Canessa had not been afraid to look down as they climbed the mountain, and each time he did so he saw that line in the far distance grow more distinct and more like a road. As they sat down in the sleeping bag and waited for the sunset, he pointed it out to the others. "Do you see that line over there?" he said. "I think it's a road."

"I can't see anything," said Nando, who was nearsighted. "But whatever it is, it can't be a road because we're facing due east and Chile's to the west."

"I know Chile's to the west," said Canessa, "but I still say that it's a road. And there's no snow down there. Look, Tintin, you can see it, can't you?"

Vizintín's eyesight was not much better than Parrado's. He gazed into the distance with his small eyes. "I can just see a line, yes," he said, "but I couldn't say if it was a road or not."

"It can't be a road," said Parrado.

"There might be a mine," said Canessa. "There are copper mines right in the middle of the cordillera."

"How do you know?" asked Parrado.

"I read about it somewhere."

"It's more likely a geological fault."

There was a pause. Then Canessa said, "I think we should go back."

"Go back?" Parrado repeated.

"Yes," said Canessa. "Go back. This mountain's much too high. We'll never reach the top. With every step we take we risk our lives. . . . It's madness to go on."

"And what do we do if we go back?" asked Parrado.

"Go to that road."

"And what if the road isn't a road?"

"Look," said Canessa, "my eyesight is better than yours, and I say it's a road."

"It might be a road," said Parrado, "and it might not; but there's one thing we know for certain. To the west is Chile. If we keep going to the west, we're sure to come to Chile."

"If we keep going to the west, we're sure to break our necks."

Parrado sighed.

"Well, I'm going back anyway," said Canessa.

"And I'm going on," said Parrado. "If you walk to that road and find it isn't a road, then it'll be too late to try this way again. They're already short of food down there. There won't be enough for another expedition like this so we'll all be losers; we'll all stay up here in the cordillera."

They slept that night with their differences unresolved. At one point Vizintín was waked by lightning in the distance and he woke Canessa, fearing that a storm was about to break over them. But the night was still clear, there was no wind, and the two boys went back to sleep again.

The night did not weaken Parrado's determination; as soon as it was light he prepared himself to continue the climb. Canessa, however, seemed less sure that he was going to return to the Fairchild, so he made the suggestion that Parrado and Vizintín leave their knapsacks with him and climb a little farther up the mountain to see if they came to the top. Parrado accepted this idea and set off at once, with Vizintín behind him, but in his impatience to reach the summit Parrado climbed quickly and Vizintín was soon left behind.

The ascent had become exceptionally difficult. The wall of snow was almost vertical and Parrado could only proceed by digging steps for his hands and feet, which Vizintín used as he followed him. If he had slipped he would have fallen for many hundreds of feet, but this did not dismay him; the surface of the snow was so steep, and the sky above it so blue, that he knew he was approaching the summit. He was driven on by all the excitement of a mountaineer whose triumph is at hand and by his intense anxiety to see what was on the other side. As he climbed he told himself, "I'm going to see a valley, I'm going to see a river, I'm going to see green grass and trees—" and then suddenly the sheer face was no longer so steep. It fell sharply to a slight incline and then flattened out onto a level surface of some twelve feet wide before falling away on the other side. He was at the top of the mountain.

Parrado's joy at having made it lasted for only the few seconds it took him to scramble to his feet; the view before him was not of green valleys running down toward the Pacific Ocean but an endless expanse of snow-covered mountains. From where he stood, nothing blocked his view of vast cordillera, and for the first time Parrado felt that they were finished. He sank to his knees and wanted to curse and cry to heaven at the injustice, but no sound came from his mouth and as he looked up again, panting from his recent exertion in the thin air of the mountain, his momentary despair was replaced once again by a certain elation at what he had done. It was true that the view before him was of mountains, their peaks standing in ranks to the far horizon, but the very fact that he was above them showed that he had climbed one of the highest mountains in the Andes. I've climbed this mountain, he thought to himself, and I shall call it Mount Seler after my father.

He had with him the lipstick he used for his chapped lips and an extra plastic bag; he wrote the name "Seler" on the plastic bag with the lipstick and placed it under a stone on the summit. He then sat back to admire the view.

As he studied the mountains spread out before him he came to notice that due west, to the far left of the panorama, there were two mountains whose peaks were not covered with snow. "The cordillera must end somewhere," he said to himself, "so perhaps those two are in Chile." The truth was, of course, that he knew nothing about the cordillera, but this idea renewed his optimism, and when he heard Vizintín calling him from below, he shouted down to him in a buoyant tone of voice, "Go back and fetch Muscles. Tell him it's all going to be all right. Tell him to come up and see for himself!" And seeing that Vizintín had heard him and was climbing down again, Parrado returned to admiring the view from the top of Mount Seler.

When the two others set off for the summit, Canessa had sat back with the knapsacks and watched his road as it changed color in the changing light. The more he stared at it the more convinced he became that it was a road, but then in two hours Vizintín returned with the news that Parrado had reached the top and wanted Canessa to join him.

"Are you sure he's at the top?"

"Yes, quite sure."

"Did you get there?"

"No, but Nando says it's marvelous. He says everything's going to be all right."

Reluctantly Canessa got to his feet and clambered up the side of the mountain. He had left his knapsack with Vizintín, but still it took him an hour longer than it had taken Parrado. He followed the steps they had cut into the snow, and as he approached the summit he called out for his friend. He heard Parrado shout back and followed his directions until he too stood on the top of the mountain.

The effect of what he saw was the same on Canessa as it had been on Parrado. He looked aghast at the endless mountains stretching away to the west. "But we've had it," he said. "We've absolutely had it. There isn't a chance in hell of getting through all that."

"But look," said Parrado. "Look there to the west. Don't you see? To the left? Two mountains without any snow?"

"Do you mean those tits?"

"The tits. Yes."

"But they're miles away. It'll take us fifty days to get to them."

"Fifty days? Do you think so? But look there." Parrado pointed into the middle distance. "If we go down this mountain and along that valley it leads to that sort of Y. Now, one branch of the Y must lead to the tits."

Canessa followed the line of Parrado's arm, saw the valley, and saw the Y. "Maybe," he said. "But it'd still take us fifty days, and we've only enough food for ten."

"I know," said Parrado. "But I've thought of something. Why don't we send Tintin back?"

"I'm not sure he'd want to go."

"He'll go if we tell him to. Then we can keep his food. If we ration it out carefully, it should last us for twenty days."

"And after that?"

"After that we'll find some."

"I don't know," said Canessa. "I think I'd rather go back and look for that road."

"Then go back," said Parrado sharply. "Go back and find your road. But I'm going on to Chile."

They retraced their steps down the mountain, reaching Vizintín and the knapsacks around five in the afternoon. While they were away Vizintín had melted some snow, so they were able to quench their thirst before eating some more meat. As they were eating, Canessa turned to Vizintín and said, in the most casual tone of voice he could muster, "Hey, Tintin, Nando thinks it might be best if you went back to the plane. You see, it would give us more food."

"Go back?" said Vizintín his face lighting up. "Sure. If you think so." And before either of the other two could say anything he had picked up his knapsack and was about to strap it to his back.

"Not tonight," said Canessa. "Tomorrow morning will do."

"Tomorrow morning?" said Vizintín. "Okay. Fine."

"You don't mind?"

"Mind? No. Anything you say."

"And when you get back," said Canessa, "tell the others that we've gone west. And if the plane spots you and you get rescued, please don't forget about us."

Canessa lay awake that night, by no means sure in his own mind that he would be going on with Parrado rather than returning with Vizintín. He continued to discuss the matter with Parrado under the stars, and Vizintín went to sleep to the sound of their arguing voices. But next morning, when they awoke, Canessa had made up his mind. He would go on with Parrado. They therefore took the meat from Vizintín and anything else that might be useful to them (though not the revolver, which they had always considered a dead weight) and prepared to send him on his way.

"Tell me, Muscles," said Vizintín, "is there anything . . . I mean, any part of the bodies that one *shouldn't* eat?"

"Nothing," said Canessa. "Everything has got some nutritional value."

"Even the lungs?"

"Even the lungs."

Vizintín nodded. Then he looked at Canessa again. "Look," he said. "Since you're going on and I'm going back, is there anything of mine you think you might need? Don't hesitate to say, because all our lives depend on your getting through."

"Well," said Canessa, looking Vizintín up and down and eyeing his equipment. "I wouldn't mind that balaclava."

"This?" said Vizintín, handling the white wool balaclava which he had on his head. "Do you mean this?"

"Yes. That."

"I . . . er . . . do you think you really need it?"

"Tintin, would I ask for it if I thought I didn't need it?"

Reluctantly Vizintín stripped off and handed over his prized balaclava. "Well, good luck," he said.

"Same to you," said Parrado. "Take care going down."

"I certainly will."

"Don't forget," said Canessa. "Tell Fito we've gone west. And if they rescue you, make them come and look for us."

"Don't worry," said Vizintín. He embraced his two companions and set off down the mountain.

Eight

1

The thirteen boys who remained in the Fairchild had watched the progress of the three expeditionaries up the mountain through their homemade sunglasses. It was easy to follow them on the first day, but by the second they had become just specks on the snow. What depressed the spectators was the slow progress they were making. They had thought that it would only take a morning, or at most a day, to climb the summit, yet on the morning of the second day the expeditionaries were barely halfway up. By the afternoon of the second day, they reached a band of shale and disappeared from view.

At the same time, however, an airplane appeared over the tip of the mountain. The boys immediately prepared to signal to it, but no sooner had they seen it than it turned back to the west.

There was nothing more they could do for the expeditionaries but pray for them; on the other hand, there were several practical problems which they faced themselves. Chief among these was the shortage of food. Although they still had not found all the bodies around the plane, Fito decided that he would climb up the mountain in search of those bodies that had fallen out of the plane. With each day that passed, new shapes and dark patches appeared on the side of the mountain, and he felt it important, if these were bodies, to go and cover them with snow before they rotted.

Zerbino, who had been up the mountain and found the bodies more than seven weeks before, agreed to accompany Fito, and the two set out early on the morning of December 13. The surface of the snow was still hard and they made rapid progress. They were better equipped than Zerbino had been on that first expedition with Maspons and Turcatti, and both boys were now better trained. Occasionally, as they climbed higher, they would pause to rest and look down on the Fairchild beneath them and the mountains beyond and to either side. The higher they climbed, the more mountains came into view—all exceptionally high and covered with snow—and this depressed them enormously. It seemed impossible that they were, as they had hoped, in the foothills of the Andes. With such gigantic peaks all around them they could only be in the very middle of the range. What chance could Canessa, Parrado, and Vizintín have of reaching Chile through such impassable terrain? It must be many miles to the nearest human habitation, and the three expeditionaries had only taken food for ten days.

"We may have to send out another expedition," said Fito. "This time with more food than before."

"Who?" asked Zerbino.

"The two of us and Carlitos, perhaps, or Daniel."

"Perhaps the plane will find us first."

They stopped and looked down at the Fairchild. Since its roof was white the plane itself was practically invisible; what stood out most clearly from that altitude were the seats, clothes, and bones scattered around in the snow.

"We'd better leave all the bones," said Fito. "They're the only things that can be seen."

After climbing for two hours the two boys came across the first sign that they were in the area where the dead bodies might be found. It was a corduroy jacket lined with wool. Fito picked it up, shook off the snow, and put it on over his sweater.

They went on up the mountain and soon saw a body lying on its back in the snow. Fito was shocked to recognize the features of another of his cousins, Daniel Shaw. He was immediately overcome with conflicting feelings; he had found the food he had been looking for, but because the body was his cousin's he was exceptionally reluctant to use it. He turned to Zerbino. "Let's go on and see if we can find the others."

They continued trudging in the snow, which was quickly becoming mushy under their feet. When they reached the area where Zerbino seemed to remember the other bodies had lain, there was nothing to be seen but odd fragments of the plane itself. One of these was large enough to be used as a kind of sled, and Fito, realizing that his duty toward the other boys left him with no alternative, took it as he turned back toward the body of his cousin.

When they reached it again they strapped the corpse, stiff from death and the cold, onto the curved piece of metal with nylon cords that came from the luggage compartment. Fito sat on one of the cushions they had brought with them as snowshoes and tied that to the sled. Then they kicked their heels into the snow and began to move down the side of the mountain.

This improvised means of transporting the corpse turned out to be more efficient than they had supposed, and as they slid down toward the valley it gathered speed until it was moving very fast indeed. Zerbino, sitting behind the body, tried to steer the sled with his feet, but he had little control—insufficient, certainly, to maneuver it between the boulders which lay strewn over the snow. A hidden hand seemed to guide them, however, for the sled did not hit a rock, and when it came to the level of the Fairchild Fito dug his feet into the snow and the cavalcade slowly came to a stop.

Their aim had been wrong. They were on the wrong side of the valley, and by now the snow was so soft, and the two of them were so tired, that they decided to leave the body where it was, buried in snow, and come back for it the next day. As they were digging a trench with their hands they saw the figures of Eduardo, Fernández, Algorta, and Páez coming toward them.

"Are you all right?" Fernández shouted.

They did not answer.

"We saw you hurtling down the mountain at such a speed that we thought you'd killed yourselves."

Again they did not answer.

"Did you find anything?"

"Yes," said Fito. "We found Daniel."

Fernández looked at Fito but said nothing. They made their way back to the Fairchild, but next morning, when the surface of the snow was frozen hard once again, they went back to fetch the body. After they had brought it back to the plane, Fito asked the other boys if his cousin's body could be placed beside those which were to be kept until the very end, and they agreed.

Páez and Algorta went up the mountain on a separate expedition to see if they could find another body. First they found a handbag from which they took the lipstick so useful for protecting their cracked and blistered lips from the sun. The two of them started to dab it onto their faces, looking at themselves in the mirror of the powder compact. "You know what they'll think," said Carlitos, laughing at Pedro's painted face. "If they rescue us this afternoon and find us looking like this, they'll think that sexual frustration has made us into raving queens."

They climbed up on the mountain and came across a body. The skin of the face and hands that had been exposed to the sun had turned black and the eyes were missing, either because the sun had burned them out or because the condors had eaten them. By now it was hot and the snow was getting soft, so they covered the body with snow and set off back down the mountain.

The next day Algorta returned to the body with Fito and Zerbino. They started to cut it up at once, having concluded that this would be easier than taking the whole body down to the plane. They put meat and fat into rugby socks and ate what they felt they deserved for the extra expenditure of energy. Then, at half past nine in the morning, though they had not finished their work, they set off back to the plane—Fito and Zerbino with full knapsacks, Algorta carrying an arm over his shoulder and the ax tied to his belt.

When they reached the Fairchild an extraordinary sight met their eyes. The other survivors were all out of the plane, standing in the middle of the cross and staring up at the sky. Some were embracing one another; others were praying aloud to God. On the outer extremity of the cross, Pancho Delgado was on his knees, shouting, "Gaston, poor Gaston! How I wish he was here now."

Daniel Fernández stood at the center of the cross, the radio held to his ear. "They've found a cross," he said. "We've just heard on the radio that they've found a cross on something called the Santa Elena mountain."

The spirits of the three who had newly arrived with the meat soared at this news, for what other cross could they have found but their own? They thought that the Santa Elena mountain must be the mountain behind them, and for the rest of the morning they waited for the rescue they believed to be imminent, Fernández always with the radio pressed to his ear. He heard that the Chilean and Argentinian planes had joined the Uruguayan C-47 in the search and that the Argentinian authorities were investigating the cross, which was thought to be on their territory.

While Fernández was listening in this way to the radio, Methol brought out a small statue of Santa Elena which Liliana had had among her belongings. Along with some other boys he prayed to that patron saint of lost things, and many of them promised that if ever they had a daughter they would call her Elena.

All that day they waited for the helicopters, and suddenly around midday they heard them from the other side of the mountains. Once again they embraced and jumped in the air, but their celebrations were premature. No helicopters appeared in the sky. The sound they had heard degenerated into a rumble and then disappeared, leaving only the huge silence of the cordillera. What they had taken to be the sound of a helicopter had been the noise of an avalanche.

When evening came they returned to the plane, bitterly disappointed. Their thoughts became more sober. What plane had flown over them, or even over the tail, that could have seen one or the other of the crosses they had made? And if they had been found, why were there no helicopters?

The next morning—very early, and in freezing weather—the same three boys as before returned to the body on the mountain to cut away the flesh that remained before it decomposed. Again they ate the extra food they felt they deserved and filled their knapsacks until only the spine, the ribs, the feet, and the skull remained This last they split open with the ax, but the brains smelled putrid so they put them aside and set off back down the mountain.

2

On the morning of December 15, those boys who were sitting on the seats laid out in front of the plane suddenly saw something hurtling at tremendous speed down the side of the mountain. They thought at first it was a boulder dislodged by the melting snow, but as it came closer they saw it was the figure of a man—and as it came closer still they recognized this figure as Vizintín. He seemed to be falling, yet his descent was controlled, for he was sitting on a cushion, and when he drew level with the plane he dug his feet into the snow and came to a stop.

A variety of dread emotions came over the thirteen boys as they watched him plod toward them over the snow. They thought that either one or both of the two other expeditionaries must have been killed or that all three had given up and Vizintín was the first to return. Some were a little more optimistic, thinking that the expeditionaries had been seen by the airplane which had appeared over the tip of the mountain.

When Vizintín reached them he explained what had happened. "Nando and Muscles got to the top," he said. "They're going on, but they sent me back to make the food last longer."

"But what's on the other side?" they all asked, clustering around him.

"More mountains . . . mountains as far as you can see. Myself, I don't think they've got much chance."

Their spirits now sank once again. Another cherished dream—that there were green valleys on the other side of the mountain—had been pushed out by brute reality. What Vizintín said depressed them all. "The climb was hell," he said, "utter hell. It took us three days to get up there. If they have to climb another like that, I don't think they'll make it."

"How long did it take you to come down?"

Vizintín laughed. "Three quarters of an hour. Coming down's no problem. It's going up." He paused, and then added, "And the funny thing is that there's less snow to the east"—he pointed down the valley—"than there is to the west. And Muscles thought he saw a road."

"A road? Where?"

"To the east."

The Strauches shook their heads. "But that's impossible. Chile is to the west."

"Yes," went up the chant from the younger boys. "To the west is Chile . . . to the west is Chile."

At midday there was a slight dispute as to how much meat Vizintín should be given. He was given the same ration as everyone else but asked for more.

"But you're no longer an expeditionary," said Eduardo.

"I know," said Vizintín. "But I've just come back from an expedition. I need to build up my strength . . . and I haven't eaten much. I gave all my food to Nando and Muscles."

They gave him a little more, but on the understanding that in future he would be treated the same as the others. That afternoon, therefore, Vizintín went around the plane picking up all the lungs he could find and piling them on his tray. Until then they had been thrown aside (except on one occasion when Canessa had palmed them off as liver), and no one had bothered to cover them with snow; they had therefore started to rot and from the heat of the sun had formed on the outside a tough leathery skin. The others watched Vizintín gather up this load and lay it on his part of the roof of the plane.

"Are you going to eat those?" someone asked.

"Yes."

"They'll make you ill."

"No, they won't. Muscles said they would do me good."

They watched him cut pieces from the putrid lungs and eat them; and when, the following day, they saw that he was no worse for having done so, some of the other boys started to follow his example. They did so from the need of a new taste, not because food was short, for suddenly the melting had started to reveal all the bodies of those who had died in or just after the original accident. There were ten such bodies, five of which they had promised to eat only from extreme necessity. One had been largely consumed before the avalanche, but the six which remained, together with the two pilots who were still in their seats, could last the survivors for another five or six weeks. All the bodies had been perfectly preserved by the snow, and because they were the first who had died they had more and better meat on them than the bodies of those who had died in the avalanche or after.

These sudden fruits of the thaw might have tempted a less disciplined group to relax the strict rationing of the meat, but the Strauches had very much in mind that they might have to mount a second expedition and equip it with supplies for a far longer journey than had been envisaged for Canessa, Parrado, and Vizintín. They therefore dug two pits in the snow. In one they put the bodies that were to be left until last and in the other those they would use as the need arose.

It would have been possible now to avoid eating such things as rotten lungs and putrid intestines of bodies they had cut up weeks before but half the boys continued to do so because they had come to need the stronger taste. It had taken a supreme effort of will for these boys to eat human flesh at all, but once they had started and persevered, appetite had come with the eating, for the instinct to survive was a harsh tyrant which demanded not just that they eat their companions but that they get used to doing so.

Perhaps the most paradoxical in this respect was Pedro Algorta. He did not come from a ranching family like the others—he was a sensitive socialist intellectual—and it was he who had justified eating those first slivers of human flesh by comparing what they were doing to eating the body and blood of Christ in the Holy Eucharist. Yet now when they discovered once again the same carcass from whom the first meat was taken, Algorta sat down on a cushion with a knife and cut away the rotten, sodden flesh which remained around the shoulders and ribs. It was still more difficult for him and for the others to eat what was recognizably human—a hand, say, or a foot—but they did so all the same.

In the middle of the day the sun was now so hot that they could almost cook their meat on the roof of the plane. There were other consequences of rapid thaw. The level of the snow fell to the base of the fuselage, which not only made it difficult to get onto the roof but also made them fear that the plane might topple over. The melting snow also dislodged boulders higher up the mountains which came hurtling down toward them. The warmth brought more signs of outside life; a few swallows flew around the plane, and one settled on the shoulder of one of the boys, who tried to grab it but failed. On the whole, however, the waiting had a bad effect on their nerves. In their minds they dithered between hope for the success of Canessa and Parrado and more sober plans for a second expedition which was to set out on the second or third of January.

It was the moment for their scapegoat to come center stage. A tube of toothpaste had been issued to the whole group, a spot at a time, as their dessert; this tube was found squeezed empty on the floor of the plane. There was an immediate inquiry, and suspicion seemed to fall on Moncho Sabella or Pancho Delgado, because they were the only ones who had been inside the plane, but since nothing could be proved no one was directly accused. In the course of this investigation, however, it emerged that Roy Harley had among his belongings his own small tube of toothpaste. When asked to account for it, he said that he had swapped it with Delgado for seven cigarettes.

"And where did you get it, Pancho?" Delgado was asked.

"Muscles brought it up from the tail and gave it to me to give to Numa. So when Numa died . . ."

"You kept it?"

"Yes."

"Why didn't you give it back to the community?"

"Give it back? I don't know. It didn't occur to me."

The matter was discussed, and twelve good men and true found that Delgado had had no right to keep the toothpaste after Numa's death, that he therefore had no right to exchange it with Roy for seven cigarettes, that the toothpaste was forthwith confiscated by the community, and that Delgado must make amends to Roy.

Roy was held to be blameless because he had been

away at the tail for most of the time that Delgado had had the toothpaste and therefore could not have been expected to know that it was held in trust for Numa. Delgado was found guilty, but it was generally held that he had done what he did in good faith, and since he gracefully accepted the verdict of his peers and returned the cigarettes to Roy (though only four, because some of the toothpaste had been eaten), this particular incident was forgotten. The suspicion still remained, however, in the minds of some of his companions that Delgado had eaten the other tube of toothpaste; though no one accused him directly, they made angry and pointed remarks to his face.

While all of them pilfered extra pieces of meat quite openly—Inciarte when he cooked—Delgado did so in secret, and because he had less opportunity than the others he took more on each occasion. The officious Zerbino, therefore, true to his role as the detective, decided to set a trap. Daniel Fernández was cutting up a body some distance from the plane. He gave the larger pieces of meat to Zerbino, who—if he did not put them into his mouth—handed them to Delgado, who passed them on to Eduardo Strauch, where they were cut into smaller pieces. Two smaller pieces did not reach their destination. Zerbino immediately called to Fito to watch Delgado and then passed along a large piece of meat. Delgado, unaware that he was being watched, slipped it onto a tray by his seat and passed on a smaller piece to Eduardo.

Zerbino immediately pounced on him. "What's going on?" he said.

"Going on?" said Pancho.

"What's that on your tray?" said Fito.

"What are you talking about?" said Pancho. "This? This piece of meat? Oh, it's a bit that Daniel left behind this morning."

Zerbino looked at him with contempt, keeping a tight hold of his temper; then he turned his back on him and went to tell the German. Eduardo was less able to contain himself. He did not speak directly to Delgado, who still sat on a seat a few feet away, but he abused him and cursed him to the others in a voice so loud that Delgado could not help but overhear.

"What is this?" he said to Eduardo. "Are you talking about me?"

"Yes, I am," said Eduardo. "This is the seventh time that food has disappeared, and there you have it on your tray."

Delgado turned pale and said nothing, and Fito took his cousin by the arm. "Leave it, let's leave it," he said.

The German's fury abated, but the disapprobation of the cousins was a severe disadvantage in the small community in the Fairchild. A strong feeling built up against Delgado. If anything was missing, pointed remarks were made in his hearing about the "opportunist" or the "holy hand." The feeling was shared even by Algorta, who slept with him at that time, and who remembered how Delgado had warmed him after the avalanche. Though he was by no means convinced that Delgado was responsible for any of the petty thefts, he was swept up in the atmosphere of antagonism. He was even afraid that if he stuck up for Delgado he would himself be isolated by the group. Methol and Mangino would not turn against him, but the only boy who still remained Delgado's friend was Coche Inciarte, for he remembered how Pancho had lent him his coat when he was cold and had forced him to eat meat and fat when revulsion led him to starve. Yet Coche was so beloved, not only by the cousins but by all the boys, that no one would turn against him for the company he kept. This one factor kept Pancho Delgado from total isolation.

Incidents such as these with Delgado did little to improve their morale. As the days passed only bad news came over the radio. The cross that had been found on a mountain was not theirs but the work of a team of Argentinian geophysicists from Mendoza. As a result the helicopters of the SAR had been grounded once again. Only the Uruguayan C-47 was continuing the search.

Then, one afternoon, they heard the drone of its engines in the sky. Once again—as when they had heard about the cross—they were thrown into paroxysms of excitement and fell to shouting and praying until, to their horror, the sound of the plane grew fainter. Then they were absolutely silent, standing in the snow and straining their ears to catch the slightest sound. The drone of the plane grew fainter, then louder again, then fainter, then louder still. They could not see it but they deduced from the sound that it was flying over the area in parallel lines. At once they prepared all their brightest garments and—realizing that the plane would be more likely to spot movement of some kind—practiced an entire routine whereby the healthiest would run around in two circles while the lame would stand in a line waving up into the sky. So that all should know where to run and where to stand, they laid out the pattern with bones—a straight line with a circle on either side. They waited until evening, the sounds of the plane's engines getting closer all the time; and when it grew dark and there was no longer any sound from the sky, they went to bed happy to think that it would almost certainly resume its search where it had left off the day before. That night, like other nights, they prayed to be rescued, but they prayed too that the expeditionaries should get help before the plane found them. The next morning—as if in partial answer to that prayer—they heard on the radio that the Uruguayan C-47 had developed engine trouble yet again and was grounded in Santiago.

It had been a week since Canessa and Parrado had left them, and in less than a week it would be Christmas Day. The thought that they were now almost certain to spend Christmas on the mountain was deeply depressing to almost all of them. Only Pedro Algorta felt reasonably content; he looked forward to the Havana cigar they were each to have to celebrate the occasion. For the others this prospect brought their spirits to their lowest ebb. Even Fito, having climbed the mountain with Zerbino and seen what surrounded them, felt doubtful that Parrado and Canessa would get through. He discussed another expedition with Páez, Zerbino, and his cousins, but with none of the optimism and enthusiasm he had shown for the first. If their champions, the expeditionaries, had failed, what chance had they of success?

During the morning their minds were distracted from this pessimistic train of thought by the work of cutting the meat. It was after they had eaten and climbed back into the plane for a siesta that they became most depressed. They could neither work nor sleep but lay listlessly in the damp, stuffy cabin, waiting for the cool of the evening. Mangino moped for Canessa. Methol, finding for the first time the letter Liliana had written to her children, wept copiously as he read it.

At three or four in the afternoon they would come out again, and these hours before dark were the most pleasant of the day. They would sit concentrating on some little task such as scraping meat off a bone or melting snow to make water and forget for a moment where they were. Then, as the sun set behind the mountains to the west, they would climb a little way up the valley and sit on cushions to smoke their last cigarette in the evening light. At this moment of the day they were almost happy.

They would talk together about anything except their homes and their families; but on the evening of December 20, as the two Strauches and Daniel Fernández sat waiting for the cold and the dark, they could not stop themselves from thinking of the Christmases of earlier years that they had all celebrated so beautifully together. The German blood was still strong enough in all three to make the idea of those celebrations going on without them particularly intolerable, and for the first time in many days hot tears began to roll down the cheeks not just of Eduardo and Daniel but of Fito as well.

Nine

1

At midday on December 12 the C-47 arrived at last at the airport of Los Cerrillos in Santiago. Páez Vilaró and his companions went to meet the pilot, who told them that he had had further trouble with the engines while flying over the Andes. It seemed that the intense cold at high altitudes affected the carburetors, and the pilot immediately arranged to have the engines checked and repaired. To assuage their impatience, the civilians tried to hire a helicopter from the "Helicopservices" which had helped them before when they were in Talca. It proved impossible. Word had got around that the search was to be in the high peaks of the central Andes, which was no place for a small bubble helicopter.

At six the next morning the C-47 was ready to fly its first mission, and it took off with Nicolich and Rodríguez Escalada on board to fly over the area of Planchon. Páez Vilaró, meanwhile, made his way south. He wanted to enlist once again the help of his friends in the Radio Club of Talca and the Aero Club of San Fernando. His tactical objective was to arrange for landing rights and refueling facilities for the C-47 at these small provincial airfields. Strategically, he wanted to stimulate interest once again in the search for the Fairchild.

On the following day—December 14—Juan Carlos Canessa and Roy Harley traveled to Curicó. Their task was to try and find the miner, Camilo Figueroa. Since his original statement soon after the disappearance of the Fairchild that he had seen the plane fall out of the skies in flames and disappear into the side of a mountain, he had vanished from sight. They talked to his brother but could get no indication whatsoever as to where Camilo might be. On the other hand they were introduced to a great friend of the missing miner, Diego Rivera, who was a miner himself and secretary of the small cooperative of miners to which Figueroa belonged. Rivera and his wife had both been in the Teno valley on October 13 and said that they had heard the engines of the Fairchild but had not seen it because of the snow that was falling at the time. Figueroa, they said, had been working closer to the point where the airplane crashed; he had seen it fly from Planchon toward Santiago and then disappear behind the Gamboa and Colorado mountains.

Canessa and Harley were encouraged by this, for it would confirm everyone's estimation that the plane must lie somewhere in the area of the Tinguiririca volcano. They immediately went to the house of one of the loyal radio hams, who put them in touch with Santiago. They talked to Nicolich, who had once again been up in the C-47, and were about to give him the information they had gleaned from Rivera when he interrupted them with the news that a cross had been seen in the snow on the side of the Santa Elena mountain.

The discovery of a cross, incontrovertibly made by men, on the slope of one of the highest mountains in the Andes range had a devastating effect in all three of the southernmost countries of the South American continent. Once again the newspapers carried headlines about the fate of the Uruguayan Fairchild; once again those parents in Montevideo who had long since despaired allowed themselves to hope; once again the air forces of Chile and Argentina began to search for survivors. And this time sorties were made not just from Santiago but from Mendoza, because the cross had been on the Argentine side of the border.

To the fathers who were in Chile—Páez Vilaró, Canessa, Harley, and Nicolich—these flights were not enough. They had seen the cross and wanted to go to it at once. For this they needed a helicopter capable of flying to such heights, but still the Chilean SAR refused to make helicopters available until there was positive evidence of human life.

Canessa and Harley would not accept this decision, and armed with a photograph of the cross they went to demand an interview with the President of Chile, Salvador Allende. They were told that Allende himself could not see them—he was resting after a tour of the Soviet Union—but through an aide he promised the Uruguayans the use of his own presidential helicopter for the following day.

It was not to be; before Allende's helicopter could come to their assistance it broke down. This further disappointment drove the fathers to distraction. After so long they had found a positive sign from their sons that they were alive, and now there was no way of going to their rescue. Immediately Páez Vilaró, Canessa, and Nicolich set off again in their own C-47 to fly over the cross once again and see if anywhere in the vicinity there was some trace of the plane itself. But when they flew into the Andes one of the engines of the C-47 failed yet again. As the four men watched the propeller come slowly to a stop—the plane lurching, righting itself, and then turning to return to Santiago—it seemed as if some malign fate were trying to frustrate them at the last moment of their momentous quest.

That morning—December 16—the Chilean Ministry of the Interior stated that the cross was a distress signal. There were, however, even among the five Uruguayans, those who had their doubts that the cross had been made by the boys because it was constructed with such geometric perfection. In Montevideo the hopes of those mothers who had had hope before were confirmed, but others were more cautious, as if afraid to let themselves believe once again that their sons might be alive. They were confused, too, because the five men in Santiago did not agree. They crowded around the radio of Rafael Ponce de León, listening to the news and talking to Páez Vilaró, Harley, Nicolich, Rodríguez Escalada—and finally to Canessa. When the last named came to the radio, Señora Nogueira took hold of the microphone and asked him what he thought, for she had great trust in his judgment.

"When I heard about the cross," Dr. Canessa said, "I wanted to parachute down. But when I saw the photograph I realized that it was much too perfect to have been made by our boys."

After that Señora Nogueira said nothing but returned

home to her husband and said, "It's not the boys' cross."

Señora Delgado, on the other hand, who had resigned herself to the fact of Pancho's death only four or five days after the accident, now came to believe once again that he was certainly alive. Her hopes were short-lived. That very afternoon—the afternoon of December 16—it was announced from Argentina that the cross had been identified as the work of a geophysical expedition from Mendoza. Twenty cones had been buried in the snow in the shape of an X. By photographing it from the air at regular intervals, the scientists could gauge the speed with which the snow was melting in the mountains and, from that, the amount of water that could be expected to pour down into the arid valleys of Argentina.

2

The effect of this news was dreadful. Señora Delgado became ill, the planes were grounded once again, and the ground patrol of the Colchagua Regiment that had been sent out in search of the cross by the commander in San Fernando, Colonel Morel, was ordered to return. But in spite of the disillusion in both countries, the five Uruguayans in Chile did not return to Montevideo. They had pledged themselves to continue the search, and this they did. On December 17, Canessa and Harley returned to Curicó to bring the miner, Diego Rivera, to Santiago. There he gave a more precise description of where the plane had been seen falling into the mountains, which once again confirmed the hypothesis that they must search in the area of the Tinguiririca volcano. The other miner, Camilo Figueroa, who had been yet nearer to the accident was still not to be found.

The next day, December 18, Páez Vilaró hired a plane to fly over the Tinguiririca, and this time he took with him not only the miner, Rivera, but Claudio Lucero, commander of the all-volunteer Chilean Andean Rescue Corps, the Cuerpo de Socorro Andino. On their second sortie they flew over a snow-covered lake leading west from the Tinguiririca, and suddenly Lucero noticed that on the lake there were the marks of human feet. The plane turned and went back for another run over the lake, flying lower so that Páez Vilaró, too, could see the footprints in the snow. "What do you think?" he asked Lucero.

"They're certainly human footprints."

"The boys?"

"Impossible. No. It must be some shepherd."

"What would he be doing walking in the snow in the middle of nowhere?"

Lucero shrugged his shoulders.

After the disappointment of the cross on the Santa Elena mountain, Páez Vilaró did not let himself believe that the footprints were those of survivors from the Fairchild. The idea which did take hold of him, however, was that they were the trail of the missing miner Figueroa, on his way to rob the dead bodies of the forty-five Uruguayans. When they landed in San Fernando he said as much to Rodríguez Escalada, adding, "Rulo, if you'll come with me, we'll get there before the robbers."

Meanwhile Dr. Canessa discussed the same prints with Lucero. "Are you sure it's not the boys?" he asked.

Lucero looked sadly at Canessa. "Doctor," he said, "it's been over two months."

Undeterred by Lucero's skepticism, Páez Vilaró went to see the Army commander, Colonel Morel, with whom by this time he had established a warm friendship. The colonel agreed to send a patrol to the area, and that afternoon he summoned an Army helicopter to take him to the valley to see for himself. He was unable to find even the footprints that Páez Vilaró and Lucero had seen, but when he returned he was still optimistic. "Listen, Carlitos," he said to Páez Vilaró. "Go home for Christmas, and while you're gone, this is what we'll do. We'll keep a patrol in the area to see if anything turns up, and in two or three days' time we'll land a group of the Rescue Corps to see what they can find. If nothing comes of all that, then you come back after Christmas and we'll start all over again."

Páez Vilaró agreed to this, as did the four other members of the group. Canessa, Harley, and Nicolich prepared to return to Montevideo the next day on the C-47, while Páez Vilaró and Rodríguez Escalada booked seats on a scheduled flight the day after.

Ten

1

After Vizintín had left them, Canessa and Parrado decided to spend the whole of that day resting near the top of the mountain. The three-day climb had left them exhausted, and they knew that they would need all their strength to reach the top again and then go down the other side. They also hoped that the airplane which had so nearly passed over them the day before might make another run in the same direction. As it was, the peace of their aerie was not disturbed. They ate their meat, melted snow for water, drank it, and thought of what lay ahead, Canessa trying to dispel his basic pessimism with thoughts such as "Qui ne risque rien, n'a rien" or "The water which falls down that side of the mountain must somehow get to the sea."

At nine in the morning on Saturday, December 16, Parrado and Canessa set off once again for the top of the mountain, Parrada first. This time they were carrying their knapsacks which, with the departure of Vizintín, were even heavier than before. It made the ascent considerably more difficult. The air at that height was rarefied; their hearts beat fast and after every three steps they would have to pause and rest, clinging to the precipitous wall of snow.

It took them three hours to reach the top. There they rested and looked over the other side for the best way down. There was considerably less snow, the valley they were making for was quite clear, but one way down looked as good as another, so they chose a path at random and set off, Parrado once again taking the lead. It was extremely difficult going, for while the sides of the mountain were not sheer, they were very steep and often made up not of solid rock but of shale. The two were attached by a long nylon luggage strap, but mostly they slid down the mountainside on their backs and bottoms—Parrado first and then Canessa—sending small avalanches of gray stone cascading down the mountain. Their knees felt weak and wobbly, yet both knew that a single slip might send them both toppling down the mountain—or they might just sprain an ankle, which in their circumstances would be as bad. Canessa began a continuous dialogue with God. He had seen the film of *Fiddler on the Roof* and remembered how Tevye had spoken to God as a friend; he now took the same tone with his Creator. "You can make it tough, God," he prayed, "but don't make it impossible."

After descending in this manner for several hundred feet, they came to a point where the side of the mountain was in the shadow of another peak and snow was still thick on the ground. The gradient was steep but the surface of the snow was solid and smooth, so Parrado decided he would toboggan down on a cushion. He untied the luggage strap, sat on one of his two cushions, stuck his aluminum pole between his legs to act as a

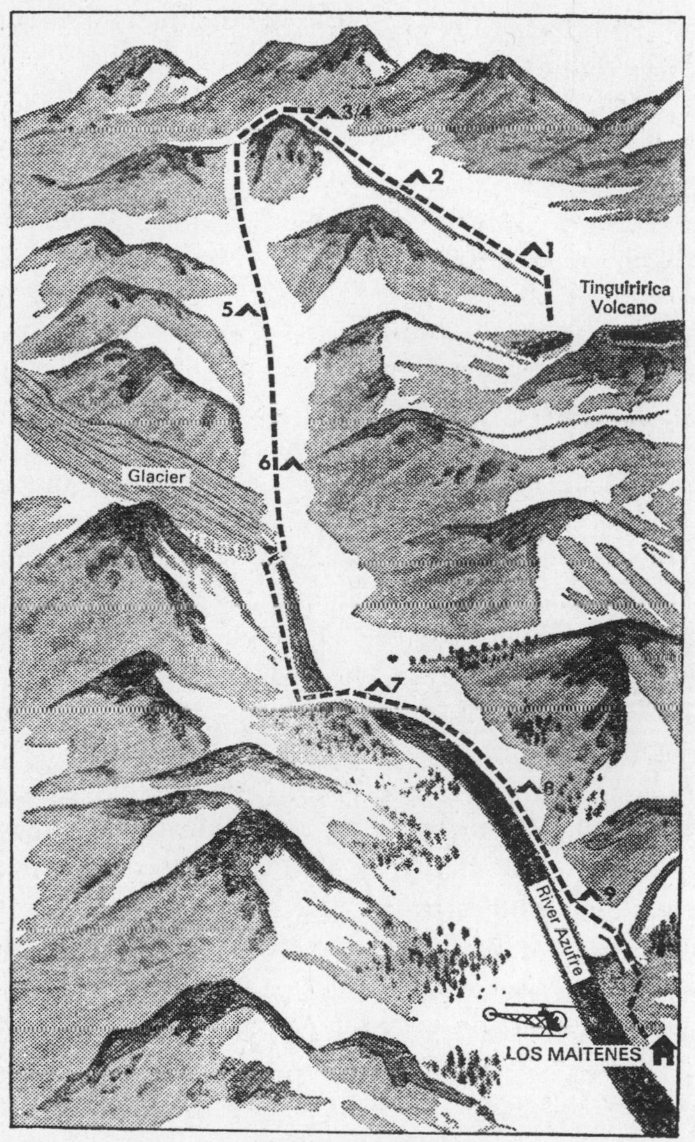

The last expedition

brake, and shoved himself off down the mountain. He immediately began to fall at a great speed, and when he dug the pole into the snow it had no effect at all. He went faster and faster, reaching a speed he estimated to be sixty miles an hour. He dug his heels into the snow but they did nothing to stop him, and he was dreadfully afraid that he would topple over and break his leg or his neck.

Suddenly in front of him he saw a white wall of snow lying right across his path. If there are boulders under that, he thought to himself, I've had it. An instant later he smashed into the wall and came to a stop. He was quite conscious and quite well. It had only been made of snow.

A moment later Canessa caught up with him. "Nando, Nando! Are you okay?" he shouted.

A tall, somewhat shaken figure climbed out of the snowdrift. "I'm okay, yes," he said. "Let's get on."

They both continued more cautiously down the side of the mountain.

At four in the afternoon they came to a large flat rock, and though they had no real idea of where they were they decided that they had better stop there and dry out their clothes before dark. They estimated that they were about two thirds of the way down the mountain. They took off their socks to dry them in the evening sun, and when the sun had set they got into the sleeping bag and slept on the rock. It was not so cold that night, but it was particularly uncomfortable.

They awoke the next morning at the first light but waited in the sleeping bag until the rays of the sun were upon them before having their breakfast of raw meat and a slug of brandy and setting off again. It was the sixth day of their journey, and at midday they reached the bottom of the mountain. They found themselves where they had planned to be—at the entrance to the valley which led to the Y. Its surface was covered with snow, which at this time of the day was mushy and deep, so they had to wear their snowshoes, but its slope was no steeper than 10 or 12 degrees. Before moving on, however, they ate their lunch. The sun was on them as they ate, and then as they walked again, which together with the exertion required to plod through the snow on sodden cushions made them both extremely hot, but they both chose to sweat under their four sweaters and four pairs of jeans rather than spare the time and energy to remove them.

Soon after they had started down the valley, the strap on Canessa's knapsack snapped and he had to stop to mend it. He was grateful for an excuse to sit down, for his strength was beginning to fail him. Whenever the intrepid Parrado looked back, he would see Canessa sitting on the snow. He would shout to him to come on, and slowly Canessa would get to his feet and plod after him. As he walked he would pray. Every step became a word of the Lord's Prayer. Parrado's mind was less on his Father in heaven than his father on earth. He knew how his father was suffering; he knew what need he had of his son. He was walking through the snow not so much to save himself as to save this man he loved so much.

With his mind on his father in this way, Parrado would draw ahead of Canessa. When he remembered his companion again, he would look around and see him several hundred yards behind. Then he would wait and, when Canessa caught up, allow him to rest for four or five minutes. On one such stop on their journey they saw to their right a small stream coming down from the side of the mountain. It was the first fresh water they had seen since Vizintín had tasted the brackish trickle which flowed over a rock on their very first expedition. From where they stood they could just see, growing around the stream, some moss, some grass, and some rushes. It was the first sign of vegetation that they had seen for sixty-five days, and Canessa, tired though he was, climbed up to that stream, picked some grass and rushes, and crammed them into his mouth. He took more of this greenery and put it into his pocket. Then both boys drank from the stream before going on their way.

As it came to the late afternoon, Canessa and Parrado began to argue as to when they should stop for the night.

"There's nowhere to sleep here," said Parrado. "No rocks, nothing. Let's go on."

"We've got to stop," Canessa replied. "I'm finished. I need to rest. And you'll kill yourself too if you don't slow down."

A momentary struggle went on in Parrado's mind between his impatience to continue and the common sense of the medical student's enjoinder to conserve his energy. It was also quite apparent that even if Parrado might survive such a forced march, Canessa would not. He therefore agreed to stop for the day, and they pitched camp on the snow. The sun had gone behind the mountains and it had started to get cold, so they climbed into their sleeping bag and warmed themselves with a drink of the brandy they carried with them. Then they lay looking down the valley which was their path to freedom, wondering what would face them the next day.

From where they lay they could see some way ahead the end of the valley which was the Y they had been making for. Both suddenly noticed that while the sun had left them at around six in the evening, it still shone on the mountain on the farther side of the Y. They watched this phenomenon with growing concentration and excitement, for since the sun set in the west, if it continued to illuminate that mountainside late into the evening it must mean that no other mountain stood in its way.

It was not until nine at night that the reddish rock streaked with snow fell into shadow. Canessa and Parrado slept that night with the firm knowledge that one arm of the Y lay open to the west.

The next morning, after their usual breakfast, they started out full of optimism, but once again Parrado drew ahead, spurred on by his curiosity to see what lay at the end of the valley. Canessa could not keep up. Little of his strength had returned with the night's rest. When Parrado stopped and turned to call to him to hurry, he shouted back that he was tired and could not go on.

"Think about something else," said Parrado. "Distract yourself from the walking."

Canessa began to imagine that he was walking down

the streets of Montevideo, window shopping, and when Parrado called to him once again to hurry, Canessa replied, "I can't hurry. I'll miss some of the store windows." Later still he distracted himself by shouting the name of a girl Parrado had once told him he liked: "Makechu . . . Makechu . . . !" Her name was lost in the snow which lay all around them, but Parrado heard and smiled and waited for his companion.

They walked on, and slowly the sound of their cushioned feet on the snow, which had been all that broke the silence, was superseded by a roaring noise which grew louder and louder as they approached the end of the valley. A panic entered the hearts of them both. What if an impassable torrent blocked their way? Parrado's impatience to see what lay ahead now took complete possession of him. His pace, already fast, quickened, and his strides grew wider over the snow.

"You'll kill yourself!" Canessa shouted after him as he drew ahead, yet he too was possessed not so much by curiosity as by dread of what they were to see. "Oh, God," he prayed once again, "by all means test us to the limit of our endurance, but please make it humanly possible to go on. Please let there be some sort of path by the river!"

Parrado walked faster and faster still. He prayed too, but above all he was obsessed with curiosity. He drew two hundred yards ahead of Canessa and then suddenly found himself at the end of the valley.

2

The view which met his eyes was of paradise. The snow stopped. From under its white shell there poured forth a torrent of gray water which flowed with tremendous force into a gorge and tumbled over boulders and stones to the west. And more beautiful still, everywhere he looked there were patches of green—moss, grass, rushes, juniper bushes, and yellow and purple flowers.

As Parrado stood there, his face wet with tears of joy, Canessa came up behind him, and he too exclaimed with happiness and delight at the sight of this blessed valley. Then both boys staggered forward off the snow and sank onto rocks by the side of the river. There, amid birds and lizards, they prayed aloud to God, thanking Him with all the fervor of their youthful hearts for having prized them from the cold and barren grip of the Andes.

For more than an hour they rested in the sun, and as if it were indeed the Garden of Eden the birds they had hardly seen for so long perched close to them on the rocks and seemed quite unalarmed by the extraordinary apparition of these two bearded, emaciated human beings, their bodies padded out with several layers of filthy clothes, their backs humped with knapsacks, their faces cracked and blistered by the sun.

They were confident now that they were saved, but they still had to press on. Canessa picked up a stone from the ground to give to Laura when he returned, and both threw aside one of their cushions, keeping only one each to sleep on. Then they started down the right-hand side of the gorge. Though there was no snow, the going was not easy. They had to walk on rough rocks and climb over boulders the size of armchairs. At midday they stopped to eat; then they went on. It was not until they had walked for another hour that Canessa realized that he had lost his sunglasses. He remembered immediately that he had taken them off and put them on a rock while they were eating their lunch, and loath as he was to walk back in the direction they had come, he was still more afraid that without the glasses his eyes would become as burned and blistered as his lips. Thus, while Parrado lay back to wait for him, Canessa retraced their steps to the spot where they had eaten. He reached it in less than an hour but, while recognizing the place itself, he could not remember on which among a hundred boulders he had put his glasses. He began to search and as he searched he prayed, for nowhere could he find what he was looking for. Tears of despair started to pour from his eyes; he was tired and desperate—until at last, on a tall rock whose top had hitherto been hidden from view, he saw his sunglasses.

Two hours after leaving Parrado, Canessa rejoined him, and both immediately continued their journey. A little farther on, however, and they were brought to a halt by an outcrop of rock which rose almost sheer in front of them and fell away precipitously into the river on their left. From where they stood they could see that the ground was more even on the other side of the river. Rather than scale the obstacle in front of them, therefore, they decided to ford the river. That in itself was no easy task. It was twenty-five feet wide, and the current flowed with such force that it carried huge boulders with it. Still there was a rock in the center of the stream which was large enough to withstand the current and high enough to stand out above the water. They decided that they could cross by leaping from the bank onto this rock and from the rock onto the opposite shore.

Canessa went first. He took off his clothes to keep them dry, tied a nylon luggage strap around his waist, and two other luggage straps to that. Then, while Parrado held the other end in case he fell into the river, he leaped onto the rock and then from the rock onto the other bank. Parrado, when he saw that his companion was safe, took the sleeping bag, tied it to the luggage strap, and threw it with all his force to the other bank. There Canessa untied it and sent back the strap so that their clothes, sticks, knapsacks, and shoes could be thrown across in the same way. It took great effort to throw the knapsacks that distance and the second fell short, crashing against the rocks by the side of the river. Canessa had to climb down to the water's edge to retrieve it, getting soaked by the spray, and when he unpacked it he found that the bottle of rum had been broken.

Parrado joined him, but since so many of their clothes were wet they walked only a little farther. Finding an overhanging shelf of rock, they decided to camp under it for the night. The sun still shone and they laid out their wet possessions to dry. Then they settled back on their cushions and ate some meat, watched by a large number of curious lizards.

That night was warmer than any thus far. They slept well and in the morning set out on the eighth day of their journey through the Andes. In the light of morning the view ahead—even to eyes less hungry than theirs for the fruits of nature—was of unsurpassed beauty. Though they were still in the shade of the great mountains behind them, the sun illuminated the farther reaches of the narrow valley, tingeing the green of the juniper and cactus plants with the silver and gold of mist and light. There were now trees to be seen in the distance, and in the middle of the morning Canessa thought he saw a group of cows grazing on the mountainside.

"I can see cows!" he shouted to Parrado.

"Cows?" Parrado repeated, blinking into the distance but seeing nothing because of his nearsightedness. "Are you sure they're cows?"

"They look like cows."

"Maybe they're deer . . . or tapirs."

What they saw before them had in any event so much the appearance of a mirage that no exaggerated expectations were placed on those distant beasts. But it meant that their spirits remained strong and optimistic just at the time when their bodies—above all, Canessa's—were suffering increasingly from the effort that had been demanded of them. The horizon might be green but the immediate terrain was no easier than it had ever been; they still had to leap, laden down by their packs, from one wobbling boulder to another, or stride on their frail ankles over the rocks and pebbles on the riverbank.

Then suddenly they came upon a most tangible sign of civilization, an empty soup can. It was rusty but the maker's name—Maggi—could still be read on the label. Canessa clutched it in his hand. "Look, Nando," he said. "It means people have been here."

Parrado was more cautious. "It might have fallen from a plane."

"How on earth could it have fallen from a plane? Planes don't have windows, do they?"

There was no way of telling how long the can had lain there, but the sight of it gave them hope, and as they continued down the valley there were other signs of life. They saw two hares leaping over the rocks on the other side of the river. Then they came upon some dung.

"That's cow dung," said Canessa. "I told you those were cows I saw."

"How do you know?" asked Parrado. "Any animal could have done that."

"If you knew half as much about cows," said Canessa, "as you know about cars, you'd know that that's cow dung."

Parrado shrugged his shoulders and they continued. Later they sat by the river to rest and eat some meat. They noticed, as they unpacked the rugby sock, that while their food supply remained adequate, it was beginning to suffer from the warmer temperatures. After eating a ration of two pieces, they repacked it all the same and set off yet again down the valley. The river was wider, for every now and then smaller rivulets would descend from the mountains on either side.

At one spot where the river widened they found a horseshoe. It was rusty like the soup can, so there was no knowing how long it had been there, yet it was not something that could have fallen from an airplane but incontrovertible evidence that they were approaching an inhabited area of the Andes. More evidence followed. As they rounded one of the many outcrops that jutted out into the valley, they suddenly came within a hundred yards of the cows that Canessa had seen from a distance that morning.

Even now Parrado was cautious. "Are you sure they aren't wild cows?" he asked Canessa, staring at the cows, which stared back at him.

"Wild cows? You don't get wild cows in the Andes. I tell you, Nando, that somewhere quite close to here we'll find the owner of those cows, or some man who's looking after them." And as if to prove that what he said was true, he pointed to the stumps of trees that had been axed by human hand. "Don't tell me that tapirs or wild cows cut down those trees."

Parrado could not dispute that the marks on the wood were those of an ax, and a little farther down the valley they found a shelter for cattle made of branches and brushwood which the two boys immediately recognized as excellent fuel for a fire. They decided to stop there for the night and celebrate their imminent salvation by feasting on the meat they had left.

"After all," said Canessa. "It's going rotten. And we're sure to find some sort of shepherd or farmer in the morning. Tomorrow night, I promise you, Nando, we'll be sleeping in a house."

They took off their knapsacks, unpacked the meat, and lit a fire. Then they roasted ten pieces each, ate until their stomachs would not take any more, and lay in their sleeping bag, waiting for the sun to set.

Now that rescue seemed so certain, they allowed themselves to think of things that until then had been too painful to contemplate. Canessa told Parrado about Laura Surraco and described lunch at her house on a Sunday; Parrado in turn told Roberto about the girls he had known before the crash and how he envied him his steady girlfriend.

The fire died down. The sun set. And with these pleasant thoughts in their minds, the two bloated boys fell asleep.

3

When they awoke next morning the cows had disappeared. This did not alarm them. They discarded what they thought they would never need again—the hammer, the sleeping bag, a pair of extra shoes, and a layer of clothes. With their loads lightened they set off again, expecting to find around every outcrop of rock the house of a Chilean peasant. As the morning wore on, however, the valley continued much as it had been. Indeed, there were not even those signs of man such as the soup can and the horseshoe which had so encouraged them the day before, and Parrado began to chide Canessa for his optimism. "So you know so much about the country, do you? So I'm just a poor fool who only knows about cars and motorbikes? Well, at least I wasn't so sure there was a farmhouse around the next corner. . . . Now we've eaten half the meat and thrown away the sleeping bag."

"The meat's gone bad anyway," said Canessa, his temper not improved by the first uncomfortable feelings of an attack of diarrhea. He was also increasingly exhausted. His whole body ached, and each step he took added to his agony. All his will had to be used to put one foot in front of the other—and when he stopped or fell behind, Parrado's curses and insults would urge him on again.

Late in the morning they came to a particularly difficult outcrop of rock where there was a choice between a shorter but more perilous route near the river or a longer but safer path over the top of the promontory. Parrado, who walked ahead of Canessa, took the more prudent route and began to clamber up the rock, but Canessa felt too tired to afford such caution, and when he came to the same spot he set off at a lower level where the ground fell away steeply toward the river below.

When he had climbed only halfway around the outcrop, stepping from foothold to foothold or moving along ledges of rock, the threatening storm broke in his bowels; his stomach was suddenly churned with the dreadful discomfort of acute diarrhea. So sharp were the waves of this disorder that Canessa was forced to find a level piece of rock, remove his three pairs of trousers, and crouch in an attempt to relieve it. In normal circumstances this would not have taken long, but here the one irregularity had been preceded by its opposite. The explosive excreta were firmly plugged into his lower colon by the rock-hard by-products of his earlier constipation, and it was only by removing these with his hands that Canessa was able to discharge what was playing such havoc in his stomach.

Parrado, meanwhile, had reached the other side and began to be alarmed and annoyed that there was no sign of his companion. He shouted for him and heard muffled words in reply. He started to curse Canessa for the delay and continued to do so until the thin, miserable figure came into view along the steep edge of the river-bank.

"Where the hell have you been?" asked Parrado.

"I had diarrhea. I feel terrible."

"Well, listen. There's some sort of track here which goes along the side of the river. If we follow that we're sure to get somewhere."

"I can't go on," said Canessa, sinking onto the ground.

"You must go on. Do you see that plateau?" He pointed down the valley to a raised piece of land. "We've got to get there by tonight."

"I can't," said Canessa. "I'm too tired. I can't walk any more."

"Don't be so stupid. You can't give up just when we're getting somewhere."

"I tell you I've got diarrhea."

Parrado flushed with irritation and impatience. "You're always ill. Look, I'll take your knapsack so now you won't have any more excuses." With that he picked up Canessa's load and set off with both packs on his back. "And if you want anything to eat," he shouted back at Canessa, "you'd better come on, because now I've got all the meat."

Canessa stumbled after him—wretched and lame—and inwardly he too was furious, not so much with Parrado for scoffing at his illness as with himself for his weakness.

It was easier to walk on the track, and every now and then they were encouraged by signs of horse dung. The symptoms of diarrhea abated in Canessa, and both boys fell into a rhythm of walking. Before them lay the plateau, and as the afternoon progressed it drew nearer. Now that they were out of the snow, distance was easier to assess. By late afternoon they had reached the escarpment which led up to it, and the promise of rest at the top gave Canessa the extra strength to follow the steep path up onto the plateau.

The first thing they saw was a corral with stone walls and a gate. In the middle there was a post driven into the ground which was used for tying horses. The ground of the enclosure had been freshly broken by horses' hooves, and both boys felt all their optimism return, but Canessa's physical condition had so deteriorated that it could not be restored by such a simple tonic as renewed hope. He staggered as he walked, and had to lean on Parrado's arm, and when they came to a small copse of trees they both agreed that they would stay there the night. It was in the minds of both of them that Canessa might have to stay longer.

While Parrado went in search of wood for a fire, and to see if by any chance there was some human habitation quite near to where they were, Canessa lay back under the trees. The ground was covered with fresh grass, the mountains rose up behind them, and the sound of the river could be heard from several hundred yards away where it crashed down through the gorge. Exhausted though he was—tired in every limb, almost to the point of extinction—the beauty of the spot was not lost on Canessa. He looked languidly at the juniper and the wild flowers, and his thoughts turned to his horse and his dog and the countryside of Uruguay.

He looked up, then, and saw Parrado returning

toward him, his tall figure bowed with anxiety. Canessa lifted himself up on his elbow. "What's it like?" he asked.

Parrado shook his head. "Not so good. There's another river which joins this one. It cuts right across our path, and I don't see how we can cross either."

Canessa sank back and Parrado sat down beside him. "But I saw two horses and two cows," he said.

"On this side of the river?"

"Yes. On this side of the river." He hesitated and then added, "Do you know how to kill a cow?"

"Kill a cow?"

"The meat's rotten. We'll need more food."

"I don't know how to kill a cow," said Canessa.

"Well, I've got an idea," said Parrado, leaning forward with a kind of earnest enthusiasm. "I know now that they sleep under some trees. Tomorrow while they're grazing I'll climb one of the trees with a rock, and when they come back at night I'll drop the rock on the head of one of the cows."

Canessa laughed. "You'll never kill a cow like that."

"Why not?"

"You'd never get up a tree with a big enough rock . . . and anyway, they might not sleep in the same place."

Parrado thought to himself in silence. Then suddenly his face lit up with another idea. "I know what," he said. "We'll take some branches and make them into spears."

Canessa shook his head.

"Or hit them on the head?"

"No. You'll never get one like that."

"Then what do you suggest?"

Canessa shrugged his shoulders.

"Come and see for yourself. They're just lying there." Parrado hesitated again. "There are the horses, though. Do you think they might go for us?"

"Of course not."

"What do you think, then?"

"Well, first of all I think that killing a cow won't make the owner of the cow inclined to help us."

"That's true."

"It might be better if we could *milk* a cow."

"But you've got to catch it to milk it."

"I know." Canessa pondered this question. "I know what," he said. "We could lasso a calf with the luggage straps and tie it to a tree. Then, when the mother comes to it, we could grab her."

"Wouldn't she get away?"

"Not if we tied her with a strap."

"But how would we take the milk with us?"

"I don't know."

"We'd have to have meat."

"Then we could kill the cow but first cut its tendons so it couldn't escape."

"And what about the owner?"

"We'd only do that if there's no one around."

"Okay." Parrado got to his feet.

"But for God's sake let's wait until tomorrow," said Canessa. "I really can't do anything else tonight."

Parrado looked down at him and saw that what he said was quite true. "Let's light a fire, anyway," he said. "Then if anyone's around there'll be more chance that they'll see us."

Parrado wandered away from Canessa in search of

broken branches and brushwood. Canessa lay back again and looked vacantly toward the other side of the river. The setting sun gave long shadows to the trees and boulders at the foot of the mountain which made them seem to move and change shape. Then suddenly, from out of these shadows, there came a moving shape, large enough to be a man on a horse. Canessa immediately tried to get to his feet, but even in his excitement his legs would hardly move, so he shouted to Parrado, "Nando, Nando! Look, there's a man, a man on a horse! I think I saw a man on a horse!"

Parrado looked up and then in the direction that Canessa indicated, but he was so nearsighted that he could not see anything.

"Where?" he shouted back. "I can't see him."

"Go, quick! Run! On the other side of the river!" shrieked Canessa in his high-pitched voice. And as Parrado started to run toward the river, he too began to scramble and crawl over the grass and stones toward the horseman three or four hundred yards, away. Every now and then he stopped and looked up to see Parrado running in the wrong direction. "No, Nando!" he would shout. "To the right, to the right!" And hearing him, Parrado would change course and run blindly on—for he still could not see anything on the other side of the river. Moreover, their shouting and waving arms had startled the cows, which had got up and were standing between Parrado and the river. They stared at him with flared nostrils, and the brave Parrado was not so brave that he did not make a slight detour to avoid them. In this way he and Canessa reached the edge of the gorge almost simultaneously.

"Where?" said Parrado. "Where's the man on the horse?"

To his great dismay, when Canessa looked over the roaring torrent to the spot where he had seen the rider, he saw only a tall rock and its lengthening shadow.

"I'm sure it was a man," he said. "I swear I saw it. A man on a horse."

Parrado shook his head. "There's no one there now."

"I know," said Canessa, sinking onto the ground and lowering his head in disappointment.

"Come on," said Parrado, taking his companion by the arm. "We'd better get back and light a fire before it gets dark."

They had both stood and faced back toward their camp when suddenly, over the splashing thunder of the river, they heard the sound of a human cry. They turned and there, on the other bank, they saw not one but three men on horses. They were staring at them while herding the three cows along a narrow path which ran between the river and the mountain.

Immediately the two boys began to wave their arms and shout at them, and the three horsemen seemed to notice them, but the noise of the river was such that their words did not seem to carry to the farther bank. Indeed, the interest the horsemen seemed to show in them was almost cursory, and it began to look as if they would ride on without reacting in any particular way to the two Uruguayans.

Parrado and Canessa became more hectic in their gestures and shouted yet louder that they were survivors from the Uruguayan airplane which had crashed in the Andes. "Help us!" they shouted. "Help us!" And while

Canessa's voice rose to a new pitch because he thought a high voice would carry farther, Parrado sank to his knees and joined his hands in a gesture of supplication.

The horsemen hesitated. One of them reined in his horse and shouted some words across the gorge, the only one of which they could decipher was "tomorrow." Then the three rode on, herding the cows in front of them.

Parrado and Canessa stumbled back to their camp. Parrado was exhausted too, and Canessa could not walk unaided. The one word they had heard, however, was enough to give them enormous hope. At last they had made contact with other men.

Although both boys were tired, they agreed that they would take turns keeping watch for two hours at a time and keep the fire alight. But in spite of their exhaustion they found it difficult to sleep. They were too excited. Then toward dawn Parrado dropped off and slept beyond the two hours they had agreed. Canessa let him sleep, for he knew that he could not walk anymore and that Parrado would need all the strength he could muster for the next day.

4

The sun rose on the tenth day of their journey through the Andes. At six both boys were awake, and looking across to the other side of the river they saw the smoke of a fire and a man standing beside it. Next to him were two other men, both still sitting on their horses. As soon as he saw them, Parrado ran once again toward the edge of the gorge. He was then close enough to the man on the other side to understand his gestures, which directed Parrado to climb down the side of the gorge to the edge of the river. This he did, and the peasant did the same until they were only separated by the thirty-five yards of the torrent itself. Though they were now closer, the noise of the cascading water was even louder than before and there was no question of speaking to one another, but the peasant, who had a round, cunning face crowned with a straw hat, had come prepared. He took a piece of paper, wrote on it, wrapped it around a stone, and threw it across the river.

Parrado stumbled over the rocks, picked up this missive, and unwrapped it. There he read:

> There is a man coming later that I told him to go. Tell me what you want.

Parrado immediately felt in his pockets for a pencil or pen but found that he only had a stick of lipstick. He therefore gestured to the opposite bank that he had nothing with which to write; whereupon the peasant took his own ballpoint pen, wrapped it with a stone in a blue and white checked handkerchief, and threw it across the river.

When Parrado had this he sat down and feverishly wrote the following message:

> I come from a plane that fell in the mountains. I am Uruguayan. We have been walking for ten days. I have a friend up there who is injured. In the plane there are still fourteen injured people. We have to get out of here quickly and we don't know

> how. We don't have any food. We are weak. *When* are you going to come and fetch us? Please. We can't even walk. Where are we?

He added to this an SOS in lipstick and wrapped the piece of paper around the stone and the stone in the handkerchief. Then he threw it back over the river.

Parrado watched and prayed as the Chilean peasant unwrapped and then read the message. At last he looked up and signaled that he understood. Then he took from his pocket a piece of bread, threw it across the river, waved once again, and turned to climb back up the side of the gorge.

Parrado did the same. He reached the plateau and walked back toward Canessa, clutching the bread in his hands, a tangible sign that they had finally made contact with the outside world.

"Look," he said to Canessa when he reached him, "look what I've got."

Canessa turned his oval face toward his friend and fixed his tired eyes on the bread. "We're saved," he said.

"Yes," said Parrado, "we're saved."

He sat down and broke the bread in two. "Here," he said. "Let's have our breakfast."

"No," said Canessa. "You eat it. I've been so useless. I don't deserve it."

"Come on," said Parrado. "You may not deserve it, but you need it."

He handed the crust to Canessa, and this time Canessa accepted it. Then the two boys sat down and ate what they had been given, and never in their lives had bread tasted so good.

Two or three hours later—at around nine in the morning—they saw another man on horseback, but this time he was on their side of the river and was riding toward them. Immediately Parrado got to his feet and went to meet him.

He greeted Parrado with great reticence, concealing the extraordinary impression that must have been made on him by this tall, bearded, bedraggled figure dressed in several suits of filthy clothes. The man followed him to where Canessa lay and listened to the babble from both of them with a patient expression on his weather-beaten face. When he was allowed to speak he introduced himself as Armando Serda. He had been told that the two Uruguayans were there, but he had understood that they were much farther upstream and had intended to fetch them that afternoon. The man who had seen them had ridden off toward Puente Negro to inform the *carabineros* of his discovery.

Parrado and Canessa could see that the peasant in front of them was poor—so poor, indeed, that his clothes were in worse condition than their own—but they suspected that, though poor, this man might have what they valued at that moment more than any treasure, and sure enough, when they told Serda that they were starving, he brought some cheese out of his pocket and gave it to the boys.

So happy were they with this cheese that Parrado and Canessa did not mind that the Chilean now left them and went on up the valley to see to the cows that were grazing there and to open some sluice gates to let water into the fields.

While he was doing this, Canessa and Parrado ate the

cheese and rested. Then, before Serda returned, they took what remained of the human flesh they had brought with them and buried it under a stone, for no sooner had the bread and cheese passed their lips than some of the early revulsion they had felt returned to them.

At around eleven o'clock that morning the peasant finished his work and rejoined the two survivors. Canessa could not walk, so he was placed on Serda's horse, and the three of them set off down the valley. When they came to the tributary of the River Azufre which Parrado had thought impassable, Serda told Canessa to dismount, and while he took the horse up the mountain toward a ford, he instructed Parrado and Canessa to cross by a footbridge which Parrado had not seen the night before.

On the other side they waited for Serda and the horse, and when he had rejoined them Canessa was lifted into the saddle once again to ride farther down the valley. There, in a meadow, they came to the first human habitation they had seen since the accident. It was a modest house rebuilt every spring, with wood and bamboo walls and a roof made with tree branches, but no palace could have seemed finer. Canessa dismounted and stood with Parrado on the grass, tipsy with the smell of the wild roses which grew over the primitive portico. Their host led them into an open courtyard, seated them at a table, and introduced them to a second peasant, Enrique Gonzalez. More cheese and then fresh milk was brought to them by this man, while Armando Serda busied himself at the stove. In a short time he brought them each a plate of beans, which he refilled four times. Both boys ate as they had never eaten before, with no thought at all to the state of their stomachs. When the beans were finished they moved on to macaroni cooked with scraps of meat, and after that bread and drippings.

At first, while they ate, the two Chileans stayed timidly at the other end of the room, but Parrado and Canessa asked them to come and sit with them. The peasants did so and sat watching the two boys stuff themselves with the food they had given them. Then, when both could eat no more, they led them to a wooden hut on the other side of their cottage. It had been built for the landlord when he came to inspect his land, and in it were two comfortable beds on which Parrado and Canessa were invited to take a siesta. With repeated professions of gratitude to their shy hosts, they did so. They had hardly slept the night before and they had been walking for ten days through some of the highest mountains in the world.

It was midday on Thursday, December 21, and seventy days since the Fairchild had crashed in the Andes.

Eleven

1

The C-47 of the Uruguayan Air Force left Santiago for Montevideo at two in the afternoon of Thursday, December 21, but while flying over Curicó the pilots were informed of bad weather on the Argentine side of the Andes, and so they returned to Santiago. The three passengers—Canessa, Harley, and Nicolich—waited at the airport until five, when they were told that the weather had improved and they could leave again. The plane took off, flew south to Curicó, then east toward Planchon, but while approaching Malargüe in Argentina it once again gave the familiar lurch which went with the failure of one of the engines.

The pilots had no choice but to make an emergency landing at the airport of San Rafael, about 185 miles south of Mendoza. There, in this small Argentinian town, the fathers spent the night. The next morning they were told by mechanics at the airport that the plane could not be repaired without parts from Montevideo. At this point the three men were inclined to find some other means to continue their journey, but there was a factor which made them hesitate—the two

Uruguayan pilots who had charge of the C-47. Both men had been friends of Ferradas and Lagurara, and while they had long since lost hope for their lives, they thought that by discovering the cause of the accident they might save their honor. They were depressed, quite naturally, by the continuous breakdown of the C-47, and it was to support and encourage them that Harley and Nicolich decided to wait there until the plane was repaired. Canessa, on the other hand, had promised to get home for Christmas, and he discovered that a bus left San Rafael for Buenos Aires that evening.

While they waited, the three men decided to contact their wives through Rafael Ponce de León's network. Once again they sought out the helpful radio ham who was to be found everywhere they went. They had some difficulty tuning in to the right wavelength because there was interference from the other hams in Chile, and it was amid the whistling and crackling that the four men caught a scrap of an exchange between two hams—"incredible, but the plane's been found." No sooner had they heard this than they lost the station.

The three Uruguayans looked at one another. "It couldn't be . . . ?" one began. The others shook their heads. Their hopes had been raised only to be dashed too often before for anything to be built on such a paltry scrap as this.

A moment later they were in contact with Rafael. They told him what had happened—that the plane was grounded and they would make their way home as soon as they could. Ponce de León promised to pass this information on to their families.

The three men wandered around the warm dry

streets of San Rafael until the time came to see Canessa onto his bus. Then at eight o'clock the doctor embraced his two friends and set off for Buenos Aires.

2

That same afternoon, Páez Vilaró and Rodríguez had driven from Santiago to Pudahuel airport to catch their plane to Montevideo. When they got there they stood in line waiting to check in their baggage, but every time the line moved forward Páez Vilaró would remain where he was, permitting those behind him to pass in front.

"Aren't you coming?" Rodríguez asked him.

"I'm waiting for something," Páez Vilaró replied.

"Well, you'll miss the plane," said Rodríguez.

"You go ahead," said Páez Vilaró. "I won't be long."

Rodríguez went on through the passport and customs controls while Páez Vilaró continued to keep to the back of the line. Then, just as the last passenger had checked in and the final call had been made for the flight, a man ran up.

"Here it is," he said, furtively handing to Páez Vilaró a small poodle puppy which he had promised to bring home to his daughters for Christmas.

Knowing that it was against the regulations to take an animal on board the plane, Páez Vilaró quickly hid the puppy under his coat and put his case onto the scales. Then, with his boarding pass in his hand, he made his way through the passport control and customs barriers. No one seemed to notice the odd position of his left arm, and Páez Vilaró was just congratulating himself on his success as a smuggler when, over the airport's loudspeaker system, came the words, "This is the international police, this is the international police. Detain Carlos Páez Vilaró. Detain Carlos Páez Vilaró."

His face fell. Someone must have seen him put the dog under his coat. He turned to the policeman who stood nearest to him and said, "I am Carlos Páez Vilaró."

He was led away across the wide foyer of the airport, cursing this latest piece of ill luck, but when he reached the office of the airport police he was faced not with handcuffs but a telephone.

"What is this?" he asked.

The officer shrugged his shoulders. "I don't know . . . an urgent call for you."

Still trying to conceal the puppy, he took hold of the telephone. "This is Páez Vilaró," he said.

"Carlitos? Is that you?" It was Colonel Morel.

"Yes, it's me," said Páez Vilaró, in a tone of mild irritation. "And I appreciate you calling me like this to say good-bye, but the plane is waiting for me. . . . I'll see you after Christmas."

"Okay," said Morel. "I'm sorry to keep you. It's just that I thought that since you've been looking for those boys of yours for so long, you might like to come and see them."

Páez Vilaró said nothing. The puppy fell to the floor.

"You could also help me with this note," Morel went on. "It might be a fake, but I don't think so. It says, 'I come from a plane that fell in the mountains. I am Uruguayan.' "

Blinded by tears, Páez Vilaró rushed from the police station out onto the tarmac. The engines of the plane had already started; they were only waiting for him to climb the steps before pulling them away.

"Rulo, Rulo!" Páez Vilaró yelled. "They found them! I'm staying."

In a moment Rodríguez was at his side, and the two weeping Uruguayans fell into each other's arms, shouting to the skies, "They're alive, they're alive!" Then together they rushed back through the customs and passport controls, shouting and weeping and causing some consternation among the various officials and fellow passengers.

"What is this?" one policeman asked another, wondering whether they should interfere with this somewhat irregular behavior.

"Leave them," said the other. "It's that lunatic who's looking for his boy who went down in the plane which crashed in the cordillera."

It was only when Páez Vilaró and Rodríguez reached the taxi stand that they realized they had no Chilean money.

"Will you take us to San Fernando?" they asked the driver of the cab at the head of the line.

"I don't know," the driver replied "It's a long way."

"They've found my son. He crashed in the Andes."

"Oh, yes," said the driver, recognizing Páez Vilaró. "You're the nut, aren't you? Okay. Climb in."

"We haven't got any money."

"Don't worry." And the two men climbed into the back of the cab.

They reached San Fernando three hours later and drove straight to the headquarters of the Colchagua Regiment. There, while the taxi driver took charge of the poodle, they were met not just by Colonel Morel but by a large crowd of all the Chileans who had been helping them in the search—the radio hams who had spread the news of the note, the pilots from the Aero Club of San Fernando, the guides, the local Andinists, and the soldiers themselves, who had so often matched fruitlessly into the mountains.

When Morel could extricate Páez Vilaró from this crowd of excited well-wishers he took him to the headquarters of the carabineers and showed him the note which had been brought from Puente Negro. "What do you think?" Morel asked. "Is it genuine?"

Páez Vilaró studied it with care. At first he was inclined to think it was a fake. So often before he had received hoax telephone calls, and here he was presented with a note which was not signed. Also, the writing was exceptionally neat for someone who had been up in the Andes for seventy days.

"I don't know," he said. "It might be a fake." Then he looked at it again and thought he recognized in the lettering something that was unique to the Stella Maris College. "But I don't know," he added quickly. "It could also be written by one of the boys."

He returned with Morel to regimental headquarters. There an operative command had been formed consisting of Morel, as commander in chief, the mayor of San Fernando, the commander of the garrison, and the commander of the carabineers. Morel appointed Páez Vilaró to this committee. "You've been searching for so long," he said. "You can't just sit around now doing nothing."

3

At midnight in San Rafael, Harley and Nicolich were in touch again with Ponce de León in Montevideo. He told them at once that a note had been handed to the police in San Fernando purporting to come from a survivor from the Uruguayan plane crash. Harley and Nicolich immediately wanted to return to Chile, but at that time of night it was not clear how they could do so. They waited, fretting and frustrated, for half an hour in the home of the radio ham, who, upon hearing the news, had run off in search of some form of transport. He returned half an hour later with the mayor's car.

Without waiting to collect any of their luggage, which was locked up in the C-47 in the airport of San Rafael, the two men set off for Mendoza. They reached it at four in the morning and went straight to the military airfield. They had no money, but when they explained what had happened, the officers of the Argentine Air Force promised them a ride on the next plane that went to Chile.

For the rest of that night they sat waiting, warmed by greatcoats given to them by the two Uruguayan pilots before they left San Rafael. At eight in the morning a plane landed with a cargo of refrigerated meat bound for Santiago. Half an hour later it took off again with Harley and Nicolich on board.

4

That same morning Dr. Canessa arrived in Buenos Aires. He had spent the night sitting in the bus and thought that before continuing his journey to Montevideo he would go to the home of a friend to wash and perhaps rest a little. He left the bus station, hailed a taxi, and slumped in the back seat as it rattled and swerved along the streets of the city.

The radio was playing some music which the driver interrupted by half turning his head toward Dr. Canessa and asking, "Have you heard the news? They've found the plane."

"What plane?"

"The Uruguayan plane. The Fairchild."

Before the taxi driver could say another word, he found his passenger sitting next to him, fumbling with the radio.

"Are you sure?" asked Canessa.

"Of course I'm sure."

"Are there survivors?"

"Two boys."

"Did they give their names?"

"Their names . . . yes, I guess they did, but I didn't really take them in."

Suddenly Canessa raised his hand to silence the driver. The news was coming over the radio that two survivors from the Uruguayan Fairchild that had crashed in the Andes on October 13 had been found in a place called Los Maitenes on the River Azufre in the province of Colchagua. Their names were Fernando Parrado and Roberto Canessa.

On hearing this last word, tears poured down Dr. Canessa's cheeks, and with a cry of happiness this strong middle-aged man turned and embraced the bewildered taxi driver as he spun his car down the streets of Buenos Aires.

Twelve

1

Canessa and Parrado, the two expeditionaries, awoke from their siesta at seven in the evening. They came out of the wooden hut into the valley, lit by the mellow light of the evening, and breathed into their lungs the warm air scented with flowers and vegetation. The beds they had slept on and the smell in the air was proof to their dazed, incredulous minds that they were no longer trapped in the Andes, but all the same they went immediately along the grass path toward the peasant's hut covered with wild roses to talk to their hosts. The two boys had talked to one another for long enough; now they wanted to talk to someone else. There was also the small matter of food, for the beans, cheese, macaroni, milk, and bread and drippings had all settled nicely into their stomachs while they slept, and both now felt ready for more.

Enrique and Armando were waiting for them and with shy sympathy understood at once what the two Uruguayans required. Though their larder was now almost exhausted, they brought out more milk and cheese and then *dulce-de-leche* and instant coffee.

As Canessa and Parrado devoured this evening meal, roasting the cheese on the fire, they questioned the two peasants about the man who had gone for the police. His name, they were told, was Sergio Catalan Martínez. He was a hill farmer, and it was he who had first seen the two boys on the other side of the river the day before. He had thought they were tourists on a hunting trip—that Canessa was Parrado's wife and that the sticks they held were rifles for shooting deer.

"But are you sure he has gone to the police?"

"Yes. To the carabineers."

"How far is the nearest post?"

Enrique and Armando looked at each other uncertainly. "At Puente Negro."

"How far is that from here?"

Again the two peasants looked at one another.

"Twenty miles? Fifty miles?"

"A day, I should say," said one.

"Less than a day," said the other.

"Walking?" asked Parrado.

"Riding."

"He went on a horse?"

"Yes. On a horse."

"And how far is the nearest town?"

"San Fernando?"

"Yes. San Fernando."

"Two days, I should say," said Armando.

"Yes, two days," said Enrique.

"On horseback?"

"On horseback. Yes."

The boys' impatience was not for themselves. With their own stomachs filled, their thoughts had returned to their fourteen friends who were still trapped in the Fairchild. They thought not only of their morale but of Roy and Coche and Moncho, whose state of health had been so bad ten days before. Every extra hour that they waited could mean the difference between life and death.

Suddenly there was a shout from farther down the valley. The two boys leaped to their feet. Parrado rushed to the entrance of the cottage, and Canessa hobbled after him. There they saw running toward them, and puffing and panting as he ran, a fat carabineer with a rope over his shoulder. Close behind him came another. They reached the cottage, and still panting from his exertion the first said to the two Uruguayans, "Okay, boys, where's the plane?"

Canessa stepped forward from the veranda of the cottage.

"Well," he said, pointing up the valley. "Do you see that opening there?"

"Yes," said the carabineer.

"Well, you go up there for about fifty or sixty miles, turn right, and then go on until you come to a mountain. You'll find the plane on the other side."

The carabineer sat down.

"Is anyone else coming?" Parrado asked anxiously.

"Yes, yes," said one carabineer. "A patrol's on its way."

Shortly afterward ten mounted carabineers were to be seen riding up the valley with peaked caps and greatcoats and ropes hanging from their saddles. Behind them, also mounted, came Sergio Catalan, the man to whom Parrado had thrown the note.

Canessa and Parrado had embraced the carabineers. They went to Catalan and embraced him too. He smiled and said very little. “Thanks be to God,” he muttered, smiling all the time, his eyes flitting from side to side to conceal his timidity. And as their gratitude became more profuse he held up his hands to restrain them. “Don’t thank me,” he said. “All I did was my duty as a Chilean and a son of God.”

The captain of the carabineers questioned Parrado and Canessa about the whereabouts of the plane. He asked if they thought it could be reached on foot, but when he heard just the outline of their journey through the Andes he realized that it would not be possible. He therefore detailed two of his men to return to Puente Negro and summon a helicopter from Santiago.

The men left with two others to guide them. By now the evening light had faded, and Canessa and Parrado realized that nothing more could be done that day. They allowed themselves to forget the fourteen up in the mountains, for a time, and settled down to talk to the carabineers—to tell them the incredible story of what had happened to the Fairchild, only with one or two details left out.

It was, perhaps, the thought of what they omitted which led the two boys to look with a certain curiosity at the packs and pouches of the carabineers. Immediately the whole patrol, enlightened as to the meaning of those glances, emptied out what they had, and Canessa and Parrado began their third feast of the day, eating eggs, bread, and orange juice and cleaning out the entire supply of the platoon of carabineers as surely as they had emptied the larder of the two peasants, Enrique and Armando. After food, however, their appetite was for conversation, and the carabineers were happy to listen.

Finally, at three in the morning, the captain suggested that they all get to sleep so as to be ready for the helicopters, which could be expected soon after sunrise.

When Canessa and Parrado came out of their hut the next day they saw to their dismay that they were in the middle of a fog bank. At the cottage they found Catalan, Enrique, Armando, and the captain all looking with equal disappointment at the thick mist.

"Can they land in this?" asked Parrado.

"I shouldn't have thought so," said the captain. "They won't find us."

"Wait," said Catalan. "It's a morning mist. It won't last forever."

The two boys sat down to a breakfast prepared by Enrique and Armando. Their disappointment at the further delay in the rescue of their friends did not diminish their enjoyment of yet another taste of normal food, and they ate stale bread and drank instant coffee with great relish. As they were coming to the end of this breakfast, they heard a strange noise in the distance. It was not the sound of a machine, so it could not be the helicopters; it was more like the twittering of a menagerie. As it grew closer and louder they could make out the sounds to be the yelps and cries of a crowd of human beings.

Imagining that the inhabitants of a local village were for some strange reason marching toward them, the boys, the peasants, and the carabineers all went out of the cottage, looked down the valley in the direction of Puente Negro, and then stood stock still in astonishment at the sight which met their eyes. Approaching them along the path over the meadow came a column of men in urban clothes, panting, stumbling, bowed under the weight of briefcases and cameras of every description. From this approaching horde there came cries of "Los Maitenes?" and "The survivors, where are the survivors?" until the first to reach the cottage saw at once, from their long hair, thin faces, and beards, which were the men they had come to see.

"*El Mercurio*, Santiago," said one, pad and pencil in his hand.

"The BBC, London," said another, one hand thrusting a microphone under their noses, the other fumbling at the controls of a portable tape recorder. Suddenly they were surrounded by fifty jostling, jabbering journalists.

Canessa and Parrado were completely bowled over by this horde of reporters. With the other boys on the mountain, they had modestly imagined that their experiences would only be of interest to one or two journalists back in Montevideo. From the limited experience of their lives they had been unable to foresee the appetite for sensation which had brought this pack in taxis and private cars along the narrow road from Santiago and then made them walk for two and a half hours, loaded with film and television cameras, along a narrow, dangerous mule path.

Faced with them, however, Canessa and Parrado were quite happy to answer their questions—again omitting one or two details, notably those about what they had eaten to stay alive. In the middle of this impromptu press conference they were called by the captain of the carabineers. The fog had lifted a little but there was still no sign of the helicopters, so the captain had decided to send Parrado and Canessa down to Puente Negro on horseback. They were mounted behind two of his men and, amid the whirring and snapping of cameras and the shouts for this pose and that, they set off down the valley, but before they had gone very far they heard the rattle of approaching helicopters from down the valley. As the deafening noise came directly over and then past them, the horses reared, wheeled, and then cantered back up the valley and reached Los Maitenes just as three helicopters of the Chilean Air Force dropped out of the cloud and landed on the far side of the river.

2

When Colonel Morel had informed the SAR at Los Cerrillos in Santiago that two survivors from the Fairchild had been found in Los Maitenes there was widespread skepticism. A request for confirmation was returned to San Fernando, but meanwhile the SAR alerted the two officers of the Air Force, commanders Carlos García and Jorge Massa, who had directed the original search for the Uruguayan aircraft. It was late afternoon (of Thursday, December 21) when García, the commander of Action Group Number 10, received the news, and he too was skeptical, assuming that Catalan had stumbled upon two mountain climbers who had been searching for the Fairchild.

In any case, it was too late to do anything that day. García therefore ordered the helicopters of his group to be ready for takeoff at six in the morning and went to bed. In the middle of the night news came through to his subordinates that the two at Los Maitenes were quite probably from the Fairchild; when García was told this the next morning it gave him a considerable shock.

In view of what he had been told, García decided to command and pilot the leading helicopter himself. To the second he assigned Massa, and to the third, as an auxiliary, a Lieutenant Avila. He also decided to take two mechanics instead of copilots, an Air Force nurse, a medical orderly, and three members of the Andean Rescue Corps, including their commander, Claudio Lucero.

The weather at Los Cerrillos was appalling. It was snowing and the visibility was only three hundred feet with a blanket of thick fog lying one hundred feet above the ground. By seven o'clock there was no indication that the weather would improve, so at 7:10 the three helicopters took off for San Fernando, flying under the fog almost at ground level.

At San Fernando they landed at the barracks of the Colchagua Regiment and were met by Colonel Morel and Carlos Páez Vilaró.

"What, you again?" said García when he saw Páez Vilaró. "You don't mean to say that you're still going on about that business of the Fairchild?"

Well might he joke. It was now confirmed that some of those he had given up for lost two months before were alive. The peasant, Sergio Catalan, had said they spoke strangely, which was consistent with a Uruguayan accent, and the names of the two boys at Los Maitenes had been given as Fernando Parrado and Roberto Canessa, both of whom had been passengers on the plane.

The command committee made a plan for the rescue. The helicopters would proceed to Los Maitenes, which would be designated Camp Alpha. They would take with them Colonel Morel, a doctor, and a medical orderly.

"And you, Carlitos," said Colonel Morel. "You deserve to go."

"No," replied Páez Vilaró. "I don't want to take up an extra place. I'll wait here." Then he turned to one of the Andean Rescue Corps and said, with deep emotion, "But if my son, Carlos Miguel, is among the survivors, perhaps you would be good enough to give him this letter. And you, Morel. Take my snow boots. You may need them, and that way you'll be walking with my feet."

The helicopters took off once again. They flew to the Tinguiririca River and then followed it into the mountains. They were equipped with charts, but on their first run up the valley they missed the point when the River Azufre branched off from the Tinguiririca, so they had to return to find it. Their information was that the carabineers were about two miles from this confluence, but the visibility was so bad that García and Massa could see nothing. It came to a point where they had either to fly blind or land—and in the relatively narrow valley, they chose to do the latter. They therefore brought the helicopters down on the left-hand side of the river.

As the noise of the motors died down, they heard shouts from the opposite bank. They went to the edge of the river, where once again an officer of the carabineers threw over a message wrapped in a handkerchief, informing them that, quite accidentally, they had come to the right place, so they got back into the helicopters and flew them to the other side of the river.

It took a short time to establish that the two emaciated, bearded figures were indeed the survivors from the Fairchild. One—Canessa—was still paralyzed from exhaustion, and the doctor and his two assistants set to work to listen to his heart and massage his aching limbs. The other, Parrado, refused such medical attention and at once began to badger García and Massa to take off again for the Fairchild. García told him that because of the fog it was impossible. He questioned him, however, about the position of the Fairchild, and Parrado described their route down the mountain.

"Have you any idea how high the plane is?" asked García.

"Not really, no," said Parrado. "Pretty high, I should say. There were no trees or plants of any sort."

"What did you eat?"

"Oh, we had some cheese, and things like that."

"Can you remember if there was any reading on the plane's altimeter?"

"Yes," said Canessa. "Seven thousand feet."

"Seven thousand feet? Good. That shouldn't be too difficult. Do you think we'll find it easily?"

Canessa and Parrado looked at each other. "Not too easily," said Parrado. "It's in the snow." He hesitated, trapped between his dread of flying and his longstanding desire to go up in a helicopter. This struggle continued for some seconds, until he remembered his promise to the boys to return for the other red shoe and blurted out to García, "I'll come with you, if you like, and show you the way."

They waited for the fog to lift. Meanwhile, many of the journalists set off to return to Santiago and file their stories. Then, three hours after he had arrived, García decided that the visibility had sufficiently improved for two of the three helicopters to take off again. With them they took the mechanics, the medical orderly, and the three members of the Andean Rescue Corps—Claudio Lucero, Osvaldo Villegas, and Sergio Díaz. Behind Díaz sat Parrado, with a helmet on his head and a microphone at his mouth.

It was around 1 P.M., the worst possible time of day for flying in the Andes. Because of this, García and Massa did not think that they would evacuate the fourteen boys on that flight but would merely establish where they were. Parrado was an excellent guide. He looked down through the windows of the helicopter and recognized all those spots on the valley where they had walked, and when they came to the Y he directed García to turn to the right and follow the narrower, snow-covered valley into the mountains.

Flying was difficult by now, but García could see from his altimeter that they were approaching 7,000 feet and felt confident that he could control his craft at that height. He saw ahead of him, however, not the wreck of the Fairchild but the sheer face of an enormous mountain.

"Where now?" he asked Parrado through the intercom.

"Up there," said Parrado, pointing straight ahead.

"Where?"

"Straight ahead."

"But you can't have come down there."

"Yes, we did. It's on the other side."

García imagined that Parrado had not heard him. "You can't have come down that mountain," he repeated.

"Yes, we did," said Parrado.

"How?"

"Sliding, stumbling. . . ."

García looked ahead, then up. What Parrado told him seemed incredible, but he had no alternative but to accept it. He started to climb. Behind him came Massa in the second helicopter. As they rose the air became thinner and more turbulent; the engines were straining with the effort, and the whole helicopter began to shake and vibrate. Yet still the mountain faced them. The peak was higher yet. The altimeter showed 10,000, then 12,000, then 13,000 feet, until at 13,500 feet they reached the top. There the helicopters were hit by a strong wind from the other side which threw them back and down. García made another run, but again the helicopter was thrown back. Parrado, behind him, screamed with fear, and Díaz, who sat next to him, told García through the intercom, "Commander, we have a panic situation back here."

García was too preoccupied with flying the helicopter to pay much attention. He saw that the mountaintop was a little lower to his right, so he gave up his assault on the peak and guided the helicopter around the top of the mountain until, still shaken and buffeted by the violent currents of air, they found themselves on the other side—but the different route had disoriented Parrado. He did not know where they were, and no one could see the Fairchild. The helicopters circled, and Parrado looked desperately around for some familiar landmark which would orient him. Then suddenly across the valley he saw the peak of a mountain which he recognized, and all at once he knew where he was. "It must be down there," he said to García.

"I can't see anything," said García.

"Go down," said Parrado.

The helicopter began to descend, and as it did so the shape of the mountains and the outcrop of rocks became more familiar to Parrado until at last he saw far beneath him the tiny specks that he knew were the remains of the Fairchild.

"There they are!" he shouted to García.

"Where, where? I can't see them."

"There!" Parrado shouted. "There!"

And at last García, still wrestling with the controls of the jumping, shaking helicopter, saw what he was looking for. "All right!" he shouted. "I can see it! Now don't talk to me, don't talk to me. Let's see if we can bring it down."

Thirteen

1

The night of Wednesday, December 20, had seen the spirits of the fourteen boys left on the mountain at their lowest ebb. It was nine days since the expeditionaries had left them—six since Vizintín had come back from the top of the mountain. They all knew what rations they had taken with them. They all knew, therefore, that time was running out. Reluctantly they faced the prospect of a second expedition—and Christmas in the Andes.

That night, after he had led the rosary as usual, Carlitos Páez said a special prayer to his uncle, who had been killed when his plane had crashed some years before. The next day, the twenty-first, would be the anniversary of that accident, and he knew that his grandmother would also be praying to her son for a special favor on that day.

The next morning they listened to the news on the radio. There was no mention of any rescue. On the contrary, it was announced that the C-47 of the Uruguayan Air Force had left Chile the day before, so they set about their duties in the same pessimistic mood. At midday they ate their ration of meat and then retired into the plane to shelter from the sun.

It was as he was leaving the plane, later in the afternoon, that Carlitos had a sudden but quite definite feeling that Parrado and Canessa had been found. He took a few paces out into the snow and went around to the front of the plane, where he saw Fito crouching by the "lavatory." He lowered himself to Fito's level and said quietly, "Listen, Fito, don't tell the others, but I've a strong feeling that Nando and Muscles have got somewhere."

Fito gave up his attempt to defecate, hitched up his trousers, and walked with Carlitos a little way up the mountain. Though not superstitious, he was happy to let this premonition dispel his gloom. "Do you really think they've found someone?" he said.

"Yes," said Carlitos in his gruff, definite voice. "But don't tell the others, because I don't want to disappoint them if it isn't true."

Once again the fourteen busied themselves with their own preoccupations while the sun sank lower in the sky. Then those who still had cigarettes (their supply was almost exhausted) lit up for that last smoke of the day. The sun set. The air grew cold. Daniel Fernández and Pancho Delgado went in to prepare the cabin, and they all lined up in pairs to enter the plane for their seventieth night on the mountain.

They prayed the rosary, and Carlitos made a special mention of his uncle but said nothing of his premonition. When the rosary was over, however, Daniel Fernández suddenly said, "Gentlemen, I have a strong feeling that our two expeditionaries have made it. We'll be rescued tomorrow or the day after."

"Me too," said Carlitos. "I felt it this afternoon. Nando and Muscles have made it."

The second premonition seemed to confirm the first, and most of the fourteen went to sleep that night buoyant with hope and optimism.

The next morning, as usual, Daniel Fernández and Eduardo Strauch went out at half past seven and tuned to Montevideo to listen to the news. The first thing they heard was that two men purporting to be survivors from the Uruguayan Fairchild had been found in a remote valley of the Andes. Eduardo was about to leap up and shout out to the others when Fernández gripped his arm. "Wait," he said. "It might be a mistake. We must be sure. We can't disappoint them again." It was he who had raised their hopes when a cross had been discovered; he did not want to do it again. The two turned their attention back to the radio and began to tune into other stations, and suddenly the whole air seemed to be alive with the news of this discovery, broadcast by public stations in Argentina and Brazil and by radio hams in Chile, Argentina, and Uruguay.

Eduardo could now shout as loud as he liked, and all the boys who were out in the snow clustered around the radio to hear with their own ears the extraordinary and magnificent jabber of the radio hams. The words, once mentioned, spread over the air from one country to another until every wavelength on the continent seemed to be carrying the sensational news that two survivors from the Uruguayan plane that had crashed in the Andes ten weeks before had been found, that fourteen still remained at the scene of the crash, and that their rescue was under way.

The moment that the boys had imagined for so long had finally arrived. They waved their arms in the air, shouting to the indifferent mountain peaks which surrounded them that they were saved and thanking God both aloud and in their hearts for the happy news of that salvation. Then they made for the cigars. There would be no Christmas in the Andes; in an hour or two the helicopters would arrive and take them away. They opened the box of Romeo and Juliet Havanas and each took one and lit up, puffing the ineffable luxury of the thick smoke into the dry mountain air. Those who still had cigarettes shared them with those who wanted them, and these were lit too.

As they smoked they grew calmer. "We'll have to tidy ourselves up," said Eduardo. "Look at your hair, Carlitos. You'd better comb it."

"What about all that?" said Fernández, pointing to the bits and pieces of human bodies which lay strewn around the plane. "Don't you think we ought to bury it?"

Fito kicked at the surface of the snow with his boot. It was still frozen hard. Then he looked up at the drawn, emaciated faces of the boys around him. "We'll never be able to dig a pit while the snow's as hard as this."

"Why bother, anyway?" said Algorta.

"What happens if they take photographs?" asked Fernández.

"We'll break the cameras," said Carlitos.

"Anyway," said Eduardo. "There's no need to hide what we've done."

Algorta could not understand what they were talking about, and the others forgot about the bodies. Zerbino and Sabella began to talk about what they would do when the rescuers arrived. "I know what we'll do," said Moncho. "When we hear the helicopters we'll go into the plane and wait there. Then, when they come to find us, we'll say, 'Hello. What do you want?' "

"And when they offer me a Chilean cigarette," said Zerbino, with a laugh, "I'll say, 'No, thanks, I prefer my own.' " He held up his packet of Uruguayan cigarettes. "I'll keep some La Paz just to be able to do that."

Feeling that their rescue must now be near at hand, the boys prepared themselves for the outside world. Páez combed his hair, as Eduardo had told him, and even put on some hair oil that he found among the luggage. Sabella and Zerbino put on shirts and neckties. All fourteen tried to find clothes that were a little less filthy. For most of them this meant taking off both the outside and the inside layers of what they were wearing and keeping the clothes they had worn in the middle. They also cleaned their teeth with the last remaining toothpaste, squeezing it liberally onto toothbrushes and washing out their mouths with snow.

They were ready, but the helicopters did not come. The radio continued to broadcast the news of their rescue. There was even a prayer of thanks from a station in Chile which moved them all as they listened to it, but by midday there was no sign of rescue and the boys were in some confusion as to whether they should return to their routine or not. Many of them by now were hungry, but the meat which had been prepared the day before had been thrown aside that morning in the excitement. Now Zerbino and Daniel Fernández began to look for it again. Roy Harley, who had thought he might wait for a proper lavatory, could wait no longer and went to defecate at the front of the plane, where he was laughed at by the others, who said that his fleshless backside was like the tail of a plucked chicken.

It also became hot, and many took shelter in the plane. While they were lying there, impatient but still ecstatically happy at the prospect of their rescue, Eduardo Strauch said to the others, "Think how awful it would be if there was another avalanche now, just before they got to us."

"It couldn't happen," said Fernández. "Not when we've come as far as this."

Suddenly they heard a shout, "Look out, an avalanche!" They heard a *woosh* and saw a mass of white approaching them. For a moment they froze with fear, but when the "snow" settled they saw it was foam from the plane's fire extinguisher, and behind it was not the sober face of death but the Cheshire-cat grin of Fito Strauch.

It was not until after one o'clock that they first heard the two helicopters and then saw them fly over the tops of the mountains slightly to the northeast of where they were. The sound was not at all what they had imagined, which proved to the boys that what they saw and heard was no mirage. Those out on the snow immediately began to shout and wave, and those in the plane tumbled out again.

To their dismay the helicopters seemed unable to see them. They flew away in the wrong direction, then turned, circled, and passed over them. This happened three times before the leading helicopter, shaking and rocking in the wind, came lower and circled above them. They could just make out Parrado, who was gesturing to them, pointing to his mouth and then holding up four fingers of one hand. They could also see that others in the helicopter were filming them and taking photographs. The pilot seemed unable to land. The wind buffeted his helicopter so badly that every time he came lower the huge machine was in danger of being blown against the rock face of the nearest mountain. A smoke canister was thrown out of the helicopter, but the smoke was blown in all directions at the same time and gave no indication of the direction of the wind. Eventually, however, after about a quarter of an hour, the first helicopter came so low that one of its skis touched the snow. Two packs were thrown from the open door, followed a second later by two men.

The first of these was Andinist Sergio Díaz, the second the medical orderly. As soon as Díaz had got clear of the blades of the helicopter, he advanced on the boys with open arms and many of them fell upon him, hugging and embracing the portly university professor and pulling him over onto the snow. Not all of them greeted him in the same way. Some of them were disconcerted by this influx of strangers into their home. Pedro Algorta, seeing Fito embrace Díaz, asked him if he already knew the man.

They were also afraid that the noise of the motors might cause another avalanche, and the two boys nearest to the first helicopter ducked under the blades and looked for a way to climb on board. It was no easy matter. García did not dare land on the snow, first because of the slope and second because he knew the snow would not bear a helicopter's weight. He was therefore hovering horizontally, afraid all the time that the blades would touch the side of the mountain, and unable to turn onto an angle which would make it easier for the boys to climb in. The first to try was Fernández. He stretched up and was grabbed by Parrado, who pulled him in. The second was Mangino, who had hobbled across the snow to get into the helicopter and now managed to scramble on board.

With these two passengers, together with Parrado, Morel, and the mechanic, García considered that he had a full load and brought the helicopter up again, and then he hovered while Massa made the same maneuver, dropping two more Andinists, Lucero and Villegas, and their equipment.

While the two helicopters were changing places in this way, Díaz extricated himself from the embrace of the boys and asked for Páez. Carlitos identified himself, and Díaz gave him two letters from his father. "One is for you," he said, "and one for the whole group."

Carlitos opened them and read first that which was for them all. "Cheer up and confidence," it read. "Here I give you a helicopter as a Christmas present." The second was to himself alone. "As you can see, I never failed you. I am waiting for you with more faith in God than ever before. Mama is on her way to Chile. Your old man." He thrust the letters into his pocket and looked up to see that the second helicopter was in position. He went toward it and, with Algorta and Eduardo, climbed on board. Behind him came Inciarte, who was helped up by Díaz, and with those four Massa had his quota of passengers and rose into the air—leaving Delgado, Sabella, Francois, Vizintín, Methol, Zerbino, Harley, and Fito Strauch with the three Andinists and the medical orderly.

The ascent of the east side of the mountain was no less terrifying than the climb up the other side—so terrifying, indeed, that the new passengers began to regret leaving the comparative safety of their "home" in the Fairchild. Fernández turned to Parrado, as the whole helicopter vibrated with the effort of the engines, and asked him if this was normal.

"Oh, yes!" Parrado shouted back, but Fernández could see from his face that he was just as frightened.

Mangino turned to the soldier sitting beside him and asked if the helicopter could cope with these conditions. The mechanic reassured him, but the expression on his face was less confident than his words, and Mangino began to pray as he had never prayed before.

The problem which faced García and Massa was that the air at that height was too thin to lift the helicopters solely by the power of their engines. What they sought to do, therefore, was to use their machines as gliders, searching for a current of hot air which would carry them up a few feet and then hovering at that level until another current carried them higher still. It was a technique which demanded exceptional skill, but not more skill than they possessed, for at last they were over the top and speeding down the valley on the other side toward the Y and Los Maitenes.

It took them only fifteen minutes to reach Camp Alpha, where the six newly rescued survivors leaped out onto the green ground in an ecstasy of joy and relief. They were amazed at the sudden color all around them, intoxicated by the smell of grass and flowers. Like drunks, they embraced one another and rolled around on the ground. Eduardo lay back on a bed of grass as though it were finer than the finest satin. He turned his head and saw a daisy growing by his nose. He picked it, sniffed it, and then handed it to Carlitos, who lay beside him. Carlitos took it from him and was about to smell it too, but instead he crammed it into his mouth and ate it.

At the same time Canessa and Algorta were embracing and rolling on the ground. In his enthusiasm and excitement, Algorta gripped Canessa by the hair. There you go again, you stupid idiot, Canessa thought to himself. Making me suffer just as you did up there.

When this first wave of ebullience had broken, and the boys realized that they had survived not just the seventy-one days in the Andes but the terrifying trip in the helicopter, their thoughts turned immediately to food, and all six fell upon the hot coffee, chocolate, and cheese that had been prepared for them. At the same time the medical team examined them and found that, while all of them were suffering in one way or another from undernourishment and vitamin deficiency, none was in a critical condition.

It was possible, therefore, for the eight survivors who had already been brought down from the site of the crash to wait at Los Maitenes while the helicopters went back for the others, but García told Colonel Morel that since no one on the mountain was in imminent danger of death it would be unwise to return that evening with the conditions so exceptionally bad. Morel agreed that the second rescue should be postponed until the next day and the first eight survivors flown to San Fernando that afternoon.

2

The command headquarters at San Fernando was informed of this decision by radio and at the same time given a full list of the sixteen survivors. The radio operator who copied it down handed it over to the committee member who had taken it upon himself to type out such messages—Carlos Páez Vilaró.

Páez Vilaró would not accept it. He knew now that of the forty passengers who had set off in the Fairchild, only sixteen had survived. He did not know if his own son, Carlitos, was one of the sixteen, and when the moment came when he could have found out, his terror of the truth proved too much. Without words he pushed the piece of paper over to Colonel Morel's secretary.

The short list was soon typed, and it was not long before, once again, the sixteen names were in front of Páez Vilaró. Still he could not bring himself to look at them. He covered the list with another piece of paper. Just as he did so the telephone rang; it was Radio Carve in Montevideo.

"Have you got any news?" they asked.

"Yes," Páez Vilaró replied. "We have the names of the survivors, but I can't give them out because we don't have the authorization of the commander in chief. He's with the helicopters."

The mayor of San Fernando, hearing this, told Páez Vilaró that he would authorize him to read out the list, so still holding the telephone to his mouth, the painter slowly uncovered the first name on the list in the same way as a Uruguayan looks at his hand during a game of *truco*. "Roberto Canessa," he said, and then repeated, "Roberto Canessa." He pulled the piece of paper a quarter of an inch down the page. "Fernando Parrado," he said, "Fernando Parrado." He pulled the paper a little farther. "José Luis Inciarte . . . José Luis Inciarte." Then a little farther: "Daniel Fernández . . . Daniel Fernández." Then a little farther: "Carlos Páez . . . Carlos Páez." Whereupon tears choked his voice, and for a moment he could read no more.

The names, as he spoke them, were being broadcast directly into the homes of every Uruguayan who had turned on the radio and had tuned it to that station. Among them was Señora Nogueira. She had been waiting for news in the garden of the Ponce de Leóns but had found the atmosphere too tense and hysterical; now she was in her own kitchen, and when Páez Vilaró began to read the list of names her whole mind and body froze as if all the hope and terror of the past two months were concentrated on that moment. "Carlos Páez," he said again. Then came the other names: Mangino, Strauch, Strauch, Harley, Vizintín, Zerbino, Delgado, Algorta, Francois, Methol, and Sabella.

There were no more names. Her hopes which had risen and sunk and then risen again so many times now sank for the last time. How nice, she thought to herself, that the little Sabella survived. She did not know his family but had talked to his mother on the telephone. Then she had seemed so sad. Now she would be so happy.

* * *

Harley and Nicolich, the two fathers who had flown from Mendoza to Santiago in the refrigerated cargo plane, reached San Fernando just as final preparations were made for the arrival of the helicopters carrying the first eight survivors. The two men did not as yet know which of the boys were still alive. They pushed their way through the crowd of excited Chileans waiting outside the entrance to the barracks and joined Páez Vilaró, who stood with César Charlone, the Uruguayan chargé d'affaires, in front of three hundred soldiers of the Colchagua Regiment drawn up as on parade.

Suddenly a cry went up from the crowd—so many of those same men who with their feet, their planes, their radios, and their prayers had helped Páez Vilaró in his quixotic search. There they saw approaching the three helicopters of the Chilean Air Force. The sight of three crosses in the sky, or a host of angels, could not at that moment have been more moving or miraculous than these loud and sophisticated machines which hacked at the air, hovered, circled, and then touched the tarmac of the parade ground.

Before their motors stopped, the doors slid back and Páez Vilaró the father saw the face of Páez the son. With a cry he surged forward and would have flung himself into the swirling blades of the helicopter had not Charlone held him back. He waited, then, while Carlitos leaped down and ran toward him. After him came Parrado and he too ran toward Páez Vilaró, who, freed by his captor, went to the two boys and embraced them both at the same time.

There were no words. For the father the weeks of

stubborn lunacy had their reward in the breathing bodies he held to his own. He wept, and behind him tears poured down the cheeks of the three hundred soldiers of the Colchagua Regiment. For the son it was enough that in the solid arms of his father he was already home. The only flaw in the happiness he felt was Nicolich's frightened, expectant face, there behind his father.

He lowered his eyes. He felt in his joy as if his arm were raised to deliver a killing blow on the head of the father of his best friend, but when he looked up again he saw that Nicolich was talking to Daniel Fernández. The look on their two faces made it quite clear what was being said.

3

The hospital of St. John of God in San Fernando had been warned at six that morning by Colonel Morel to expect the survivors from the Uruguayan Fairchild. The director of the hospital, Dr. Baquedano, immediately formed a team of his most able subordinates—Drs. Ausin, Valenzuela, and Melej—to prepare for their arrival. At that time they had no way of knowing the condition the survivors would be in. All they knew was that they had been trapped high in the Andes with little or no food for more than seventy days.

Their first step was to prepare accommodation. The hospital was small and its buildings were old—a single-story construction with interior courtyards and a covered veranda around each block. But there was a wing reserved for private patients, and it was decided that this should be evacuated for the use of the survivors. While this was being done the three doctors telephoned another Dr. Valenzuela, the chief of the Intensive Care Unit in the Central Hospital in Santiago. They explained to him what treatment they had planned for the patients when they arrived, and they received his confirmation that it was correct.

The ambulances arrived with the first eight survivors at ten past three. The boys were driven into the courtyard between the main building of the hospital and the brick chapel at the side and then wheeled in on stretchers—all except for Parrado, who insisted on walking himself, pushing his way through the crowd of nurses and visitors who were watching them arrive. When he reached the entrance to the private wing, he was stopped by the policeman who had been stationed there.

"I'm sorry," he said. "You can't go in there. It's for the survivors."

"But I am a survivor," said Parrado.

The policeman looked at the tall youth in front of him, and only the beard and long, unkempt hair persuaded him that what Parrado said was true. The nurses, too, were incredulous and tried to get the expeditionary to bed, but he refused to lie down and be examined by the doctors until he had had a bath. The nurses looked bewildered and went to ask the doctors, who shrugged their shoulders and said that they might as well let him have his way. It was their first intimation that their patients were not going to behave quite as they had expected them to.

A bath was run for Parrado. He asked for some shampoo, and a nurse went to fetch her own bottle.

Then, at last, Parrado took off his stinking clothes and sank his body into the bath. He washed himself all over and then lay back in the hot water for an hour and a half. After the bath he took a shower to wash off the dirty water and then put on the white tunic that the hospital had provided. He felt magnificent, and with a benign indifference he allowed the three perplexed doctors to examine him. They could find nothing wrong with him at all.

Of course Parrado, like the other seven, was severely underweight. He weighed at least 50 pounds less than he normally did. Those who had lost least were Fernández, Páez, Algorta, and Mangino, who had lost between 30 and 33 pounds each. Canessa had lost 37½, Eduardo Strauch 44, and Inciarte 80 pounds. This enormous difference showed not only how thin they had become but how heavy they had been before; for while Fernández, Algorta, Mangino, and Eduardo Strauch had weighed between 145 and 165 pounds before the accident, Parrado and Inciarte had both weighed almost 200. It surprised the doctors that Páez, who was not a tall boy, should weigh as much as 151 pounds when he was admitted to the hospital of St. John of God in San Fernando.

Some of the boys had specific complaints which the doctors did what they could to treat. Mangino had a fractured leg; Inciarte's leg was still badly infected; Algorta had a pain in the region of his liver. Mangino also had a slight fever, high blood pressure, and an irregular pulse. Furthermore, the tests that were taken revealed a deficiency of fats, proteins, and vitamins in them all. They all also suffered from burned and blistered lips, conjunctivitis, and various skin infections.

It soon became clear to the three doctors who were examining them that these eight boys had been nourished on something more than melted snow over the past ten weeks, and as he examined his leg one of them asked Inciarte, "What was the last thing you ate?"

"Human flesh," Coche replied.

The doctor continued to treat the leg without any comment and without showing any surprise.

Fernández and Mangino both told the doctors what they had eaten up on the mountain, and again the doctors made no comment one way or the other, although they did issue strict instructions that no journalists be admitted to the hospital. Still, it did not occur to them to alter the instructions they had given for the feeding of the eight survivors. Mangino, Inciarte, and Eduardo Strauch—who were the most critically undernourished—were fed intravenously, and the others were given liquids and modest portions of specially prepared gelatin and then left to rest.

It was some time before the boys realized that this was all they were to be given for the time being. The only solid food they had seen since entering the hospital was the piece of cheese that Canessa had brought with him from Los Maitenes as a souvenir and kept on the table beside his bed. Each boy was in his separate room, but the stronger went from one to the other and it was soon agreed that they should ask the nurses for something more substantial to eat. The nurses replied to their request with the strict instructions of Drs. Melej, Ausin, and Valenzuela that no further food was to be provided.

Ruthlessness comes from necessity. Carlitos Páez had realized that he was in some sense a celebrity, and he promised the nurses all sorts of "special" autographs and souvenirs if they would only get him something solid to eat. The charming Chilean nurses were not so easily bribed. The boys therefore began to draw up a special petition of complaint against Drs. Melej, Ausin, and Valenzuela, who were, they said, starving them to death.

The doctors returned to the ward and heard the petition. They replied by explaining the dangers of eating solid food after long periods of starvation.

"But doctor," said Canessa, "I had beans and macaroni for lunch yesterday, and I'm quite all right today."

The doctors gave up. They ordered the nurses to bring a full meal for each of the eight survivors.

It was clear by now that none of their eight patients was in a critical physical condition. The doctors' concern shifted to the area of their mental health. They had noticed from the very first two symptoms among them—first the compulsion to talk and second a dread of being left alone. Coche Inciarte, who had been the first to enter the hospital, had grabbed Dr. Ausin's hand from the stretcher and gripped it until he had been laid on his bed. Thereafter he talked to everyone who came into his room; so too did the others, especially Carlitos Páez.

This behavior was not extraordinary in young men who had spent ten weeks stranded in the Andes, but taken with the knowledge, newly acquired by the doctors, that their patients were alive thanks to a diet of human flesh, it could be the first manifestation of more extreme psychotic behavior. For this reason the doctors gave instructions that no one was to be let in to see them—not even the mothers of Páez and Canessa, who had already arrived from Montevideo.

One man, however, in the crowd of people crammed into the hall of the hospital, was made an exception to their rule. This was Father Andrés Rojas, the curate in the parish church of San Fernando Rey. Like everyone else in San Fernando, be had heard of the "Christmas Miracle" (as the discovery of the survivors had now come to be called) and that afternoon had seen the helicopters fly over the town as they arrived from Los Maitenes. He had felt an immediate impulse to go to the hospital and offer what help he could, but this impulse had been countered by some reluctance to interfere with the hospital authorities. Later that afternoon, however, he had remembered some other business which gave him an excuse to go there.

He was a young man of twenty-six and had been ordained only the year before. He looked even younger than his age, being small, with dark hair, dark skin, and a boyish physiognomy. He dressed, too, not in a black clerical habit but in gray slacks and an open-necked gray shirt, so when he got to the hospital he might not have stood out from the crowd had not Drs. Ausin and Melej recognized him and invited him to visit the survivors.

He was ushered to the private wing, and there he went into the first room leading off the corridor, which belonged to Coche Inciarte. It was a good choice, for no sooner was he identified as a priest than a gush of words poured out of the stuttering Coche. He told Father Andrés about the mountain—not in the cold language of a detached observer but in lofty mystical words which more accurately conveyed what the experience had meant to him.

"It was something no one could have imagined. I used to go to mass every Sunday, and Holy Communion had become something automatic. But up there, seeing so many miracles, being so near God, almost touching Him, I learned otherwise. Now I pray to God to give me strength and stop me slipping back to what I used to be. I have learned that life is love, and that love is giving to your neighbor. The soul of a man is the best thing about him. There is nothing better than giving to a fellow human being. . . ."

As Father Andrés listened, he came to understand the exact nature of the gift to which Inciarte referred—the gift by his dead companions of their own flesh. No sooner did he realize this than the young priest reassured him that there was no sin in what he had done. "I shall be back this afternoon with Communion," he said.

"Then I should like to confess," said Coche.

"You have confessed," said the priest, "in this conversation."

When he came to see Alvaro Mangino, Father Andrés was faced with the same phenomenon—an intense desire to account for what he had done, told with a confusion of remorse and self-justification. Once again the priest reassured the worried boy; there was no sin in what he had done. He could take Communion that afternoon without confessing or, if he wished to confess, it should be to unburden himself of other sins.

There was one question, however, which Inciarte had asked him and he could not answer. Why was it that he had lived while others had died? What purpose had God in making this selection? What sense could be made of it? "None," replied Father Andrés. "There are times when the will of God cannot be understood by our human intelligence. There are things which in all humility we must accept as a mystery."

In permitting Father Andrés to visit the survivors, the doctors had chosen a most healing therapy. The decision to eat the bodies of their friends had been a severe trial for the consciences of many of the boys on the mountain. They were all Roman Catholics and were open to the judgment of their church on what they had done. Since it is the teaching of the Catholic Church that anthropophagy *in extremis* is permissible, this young priest was able not so much to forgive them as to tell them that they had done nothing wrong. This judgment, backed with all the authority of the church, assured the peace of mind at least of those who felt uncertain.

There were one or two who did not feel the need for reassurance—or did not admit to such a need—and they had little to say to Father Andrés. Algorta, for one, was not in the mood for talking to anyone—least of all to this youthful cleric with the saintly expression on his face.

4

The moment of reunion could not be postponed much longer. The parents and relatives of the survivors were not aware of the delicate matter which the doctors had to consider before admitting them, and though most were heroically patient while waiting to see their sons who had risen from the dead, others grew increasingly impatient.

Graciela Berger, Parrado's married sister, was incensed to be stopped by a policeman at the door of the private wing. "But I want to see my brother!" she shouted.

She heard from within Nando's voice, shouting, "Graciela, I'm here!" Whereupon the determined Uruguayan girl pushed her way past the policeman and into Parrado's room. No sooner did she see him than she burst into tears. The combination of emotion at seeing her brother again, together with the shock of his unkempt and emaciated appearance, made too great an assault on her self-control.

Behind her came her husband Juan, and after him the bowed, weeping figure of Seler Parrado. This poor man, who had lost all taste for life when his son, wife, and daughter had disappeared in the Andes, had had his hopes raised again by a false list which had put all three as survivors. It was now only that he had been told the truth: only Nando was alive. He was therefore torn by the conflicting sentiments of joy and sorrow—but when he saw his son and took him in his arms, the former feeling overcame the latter and the tears which flowed from his eyes were tears of joy.

Farther along the corridor, also in a room of his own, Canessa lay on his bed listening to the voices of the relatives as they entered. Suddenly he looked up and saw at the door the face of his *novia,* Laura Surraco. His first reaction was shock. He had always thought, up in the mountains, that he would see her again in Montevideo. Her presence here in Chile was somehow wrong—and when she rushed toward him and burst into tears he drew back from her.

Laura Surraco was followed by Mecha Canessa. She walked in with serenity and said, "Merry Christmas, Roberto." Then she began to cry also, as she saw the wizened face of an old man beneath the beard of her son. When Dr. Canessa entered the room, he too burst into tears, and this torrent of emotion set Roberto crying until his parents feared for his health and offered to leave him. But he would not let them go, and when everyone was calmer he began to tell them about the accident and their survival, including the fact that they had eaten human flesh. Of the three who received this information, only the father started in shock before gaining sufficient control of himself to conceal his feelings. The two women seemed so happy to have Roberto there that they hardly cared what he said. The doctor, on the other hand, knew just what horrors his son must have been through and just what trials would await him.

The reunion of Eduardo Strauch with his mother and father and his Aunt Rosina had some of the same awkwardness about it. Sarah Strauch did her best to appear calm but she was naturally nervous, and the moment of reunion with Eduardo was almost too much for her self-control. Being possessed all at once by joy at seeing him alive, shock at seeing him so wasted, and spiritual exhilaration at this answer to her prayers to the Virgin of Garabandal, she could not control the aghast expression on her face when Eduardo told her of the extremes to which they had gone to make this miracle come true.

In the same way Madelón Rodríguez gave an involuntary grimace of horror when Carlitos told her what they had eaten to stay alive. Like many of the other mothers who still had faith in their sons' survival, she had not thought, in detail, of how this miracle might be achieved; she had assumed that there would be woods to shelter them, with rabbits running over the pine needles and fish swimming in the streams. Neither she nor any of the other parents and relatives of the boys had imagined that they might be eating the bodies of the dead. It was inevitable, therefore, that the first knowledge that this was what had happened should appall them, and the boys—even in the raw state in which their ordeal had left them—retained sufficient judgment and intelligence about human behavior outside the Andes to understand that this should be so.

The exception was Algorta. He lay on his bed in his private room, without any compulsion to talk, and sent the pious priest on his way. He was visited by his father, and a woman whom he recognized as the mother of the girl he had been going to see in Santiago. He asked after her daughter, and the next morning the girl herself came to visit him. At once he felt all the affection for her that had existed before the accident. The only flaw to the reunion was the expression of horror on the mother's face when Algorta told her what he had eaten to survive. Her shock shocked him. How could she be surprised that they ate the dead when it was such a normal and obvious thing to do?

Coche Inciarte, who was the most vulnerable to the involuntary expressions of censure in others, avoided such reactions by speaking of their experiences only in the most lofty terms.

"Carlos," he said to his uncle—the first of twelve relatives to arrive in Chile—"Carlos, I'm full of God."

And his uncle replied in the same kind. "Christ wanted you to come down from the Andes, Coche, and now He is with you."

Fourteen

1

The eight survivors who were left behind watched the ascent of the two helicopters until both had disappeared over the mountain. Then Zerbino turned to Lucero, one of the three Andinists, and invited him to visit their "home"—the hulk of the Fairchild—while waiting for the helicopters to return. As they made their way toward the entrance, Lucero glanced at the fragments of human bodies which lay scattered over the snow and said, "Have condors been eating the bodies?"

"No," Zerbino replied, following the direction of his eyes. "We have."

Lucero said nothing and showed no surprise, but when he reached the Fairchild—its roof covered with strips of fat—he hesitated a moment. Then he stooped and went in. Zerbino went in with him and explained how they had lived and slept in the confined quarters for so long, and how the avalanche had swept down the mountain and killed eight of those who had survived the accident. Lucero listened with great sympathy and interest but was unable to ignore the stench which pervaded the inside of the plane. Unfortunately his host seemed unaware that there was any smell at all. The Andinist was too polite to mention it, but he returned as quickly as he could to the open air.

Meanwhile, the other visitors had been seeing to the remaining survivors—to their medical needs and to the more urgent demands of their bellies. First came steak sandwiches—slabs of cooked beef between slices of unleavened bread. Then came orange juice, lemon juice, soup (heated on the Andinists' stove), and finally some fruitcake which Díaz had brought because it was his birthday. It was a feast. The boys ate and drank with gusto and delight, scooping up the butter with their fingers.

In preparation for the return of the helicopters, the Andinists now tried to construct a landing pad in the snow. They demolished the wall at the entrance to the plane, took a large slab of plastic which had once been part of the partition between the passenger cabin and the luggage area, and laid it on the snow in as horizontal a position as possible. They used pieces of cardboard, too, in preparing for the return of the helicopters, which had been gone for some time.

They also began to take photographs of what they saw around them. On seeing this the boys became upset and asked them why this was necessary. The Andinists calmed them by saying that they were only making a record for the Chilean Army, as they were required to do, and that none of the photographs would ever be published.

The boys seemed satisfied, and in any case it was hard to be angry with their rescuers, especially with rescuers who were as kind as the four Chileans. One of them even offered Zerbino a cigarette.

"No, thanks," said Zerbino, "I prefer my own." And without a smile to betray that he had rehearsed these lines before, he lit one of his last Uruguayan cigarettes.

By around four in the afternoon it became clear that the helicopters would not be returning that night. At once the high spirits of the eight survivors gave way to the miserable thought that they would have to spend another night on the mountain. The Andinists, noticing this, did what they could to restore their morale. They lit their stove and cooked up some more soup—first chicken soup, then onion soup, then Scandinavian soup. Then Methol asked them if, by any chance, they had some maté.

"Maté? Of course we've got maté. How can you imagine that four Chileans would be without maté?"

They brewed maté, and after that they made coffee. But by this time the sun had slipped behind the mountains and it was beginning to get cold. For the boys this sudden fall in the temperature was routine. The Andinists, too, were equipped with bright-colored waterproof anoraks which protected them against the cold. The one who began to suffer was Bravo, the medical orderly, for he had jumped out of the helicopter dressed in a shirt with short sleeves and wearing only moccasin shoes, so the Andinists found him some clothes.

The four Chileans, now sitting in the plane with the boys, started to sing songs to keep up their spirits, but as the sun sank farther behind the mountain it became colder still, and the time came for them all to try and sleep. The eight boys quite naturally invited their visitors to stay with them in the plane, but their Chilean guests showed some reluctance and instead left the

Fairchild and pitched a tent in the snow some distance away. The eight boys were somewhat offended that their hospitality was spurned in this way—though some had divined by this time that the inside of the plane might not smell as sweet to others as it did to them—and they made up their minds that at least one of the Andinists should spend the night with them.

They chose Díaz because the next day was his birthday. They said that if he did not come with them they would pull out the tent pegs at midnight. Díaz gave in. While his fellow Andinists and the medical orderly retired to the tent, he helped the boys rebuild the wall at the entrance to the plane and then clambered in with them and sat in the middle of the eight stinking, emaciated, happy Uruguayans. No one slept or even tried to. Díaz talked to the boys about the life of an Andinist and told them of some of his adventures in the mountains. They in their turn told him in more detail of their ordeal. He warned them that what they had done might come as a shock to the outside world.

"But will people understand?" the boys asked him.

"Of course," he reassured them. "When the full facts are known, everyone will understand that you did what had to be done."

At midnight Díaz was forty-eight years old, and the eight Uruguayans sang "Happy birthday to you."

Early the next morning their thoughts turned to breakfast. All the food was in the tent, however, and when they went out into the snow at first light they could see no sign that the three other Chileans were awake. A chant began to sound across the valley—"We want our breakfast, we want our breakfast!"—and in a short time the sleepy faces of Lucero and Villegas appeared between the canvas flaps of the tent.

"What do you want for breakfast?" Lucero shouted.

"We had coffee yesterday," one of the boys shouted back. "Today we want tea!"

"Tea? All right."

Lucero, Villegas, and then Bravo emerged from their tent, and in a short time the eight boys were breakfasting on hot tea and dry biscuits. As they ate the Andinists described how to approach the helicopters when they arrived, for they had given up the idea of a landing pad and knew that the machines could only hover over the snow.

After their breakfast the boys prepared themselves for rescue. They straightened their clothes and combed their hair once again, and Zerbino brought out of the plane the suitcase which he and Fernández had filled with all the money and documents of the boys who had died. He also brought the tiny red shoe which made up the pair that Parrado had bought in Mendoza.

"You won't be able to take that on the helicopter," one of the Andinists said, seeing Zerbino with the suitcase.

"I must," Zerbino replied. He explained what the suitcase contained and described to Lucero where the bodies lay which had fallen out of the plane at the top of the mountain.

At about ten o'clock they heard the sound of helicopters and then saw three appear in the sky above them. In the less turbulent air of the morning, the helicopters took none of the buffeting that they had suffered the day before; all the same, they did not descend at once but circled over the wreck of the Fairchild. The boys, who were waving frantically at the machines, could see cameras and film cameras protruding from the windows. Finally these were withdrawn and the first machine came lower and lower until one of its skis was resting on the snow.

The first three boys moved forward, but it was difficult to approach the roaring machines because of the wind which came from the blades. Roy Harley was extremely weak, and he was helped toward the helicopter by Bobby Francois, but Francois himself was blown back by the gust of wind generated by the helicopter. It was only with the help of the Chileans that they climbed on board.

Finally the first helicopter rose in the air with its three passengers and the second lowered itself to the ground, the pilot trying to keep his blades from hitting the rock, the next group of survivors trying to keep the blades from cutting off their heads. Then that too was loaded; it rose and the third took its place and the last two survivors climbed on board, including Zerbino with his suitcase. There were less treacherous currents of air for the pilots to contend with that morning, and in a short time they were over the mountains and flying more easily down the valley toward the Y. Through the transparent walls of the helicopters they saw the River Azufre, its banks sprinkled with green vegetation which increased in area and density until they landed in the lush pastures of Los Maitenes.

There the boys tumbled out of the helicopters and fell onto the grass—laughing, rolling, embracing one another, and praying aloud in thanks to God. They were stunned by the verdancy all around them. Like Páez the day before, Methol picked a flower and began to eat its stalk; so ecstatic was he at this sight of trees and clover that he made up his mind to return to Uruguay and spend many months at his *estancia,* simply staring at the green landscape all around him.

García's helicopter took off once again to return for the Andinists and the medical orderly. Meanwhile an army doctor, Sanchez, examined the survivors to see if any of them required emergency treatment. His conclusion was that all were fit to travel; indeed, the eight Uruguayans were behaving less like invalids than like young men on a picnic. While some washed themselves in the stream which ran beside the house of Serda and Gonzalez, others talked to Sergio Catalan and his sons, and Fito Strauch and Gustavo Zerbino borrowed his horse to go for a ride.

They stayed at Los Maitenes for around half an hour, after which García returned with Lucero, Díaz, Villegas, and Bravo, and the whole group reembarked in the helicopters and flew on to San Fernando. Fito Strauch, Bobby Francois, Moncho Sabella, and Gustavo Zerbino were in the first helicopter, and no sooner had it touched down at the headquarters of the Colchagua Regiment than Fito's mother and father came toward their son. Their faces were masks of pain, such was their joy, and no sooner had the blades of the helicopter come to a stop and the door opened than they had their son in their arms. Rosina embraced him, and as she held him she prayed to her little Virgin of Garabandal who had wrought this miracle.

Behind the Strauches came the Zerbinos—their eyes

dry, their faces serene—to meet their husky, healthy son, and after them—as if an order of precedence had been established according to the degree of faith each parent had had in the son's survival—came the mother and father of Bobby Francois. The child whose death they had accepted ten weeks before was now with them, and the joy seemed almost to incapacitate Dr. Francois. A suave and silent man, he had grown thin since his son had last seen him, and as he walked to the helicopter he seemed bewildered and confused. For the Strauches, and even for the Zerbinos, the return of their son was a vindication of all their effort and hope. For Dr. Francois it was a resurrection which, as a man of science, he had neither asked for nor expected.

In the next helicopter came Roy Harley, Javier Methol, Antonio Vizintín, and Pancho Delgado. Of these four, only Roy found his parents waiting for him. For them too it was a reward for faith and effort, but as with all the parents their joy was mingled with the pain of compassion for what their son had suffered. They saw that the boy who had left them as a hefty rugby player was now no more than skin and bone. All the flesh had gone from his body. His eyes were sunk in their sockets; his skin was stretched over his cheekbones; his hands were like the hands of a skeleton covered with a wizened parchment. And it was not just these physical manifestations of starvation and privation which told them what he had suffered; there was also the expression in his eyes.

After their reunion with those parents who were there, the eight survivors were taken to the regimental infirmary for yet another checkup while the doctors, the

Command Committee, and the Uruguayan chargé d'affaires, César Charlone, discussed whether they should be taken to the hospital in San Fernando—as the others had been—or flown directly to Santiago. Once it was established that they were all well enough to continue their journey, the latter course was decided upon. It was now Saturday, December 23, and it was thought important for the peace of mind of the whole group that they should be reunited with their relatives for the feast of Christmas. The survivors therefore embarked in the helicopters for the third time that day. The Harleys and Señora Zerbino were also allowed to travel with their sons, and the whole party took off from the headquarters of the Colchagua Regiment and landed a short time later on the roof of the Public Assistance Hospital—the Posta Central, as it is called—in Santiago.

2

Meanwhile, in the hospital of St. John of God in San Fernando, the first group of survivors had slept their first night in a bed for seventy-one days. It was not easy for them to accustom themselves to comfort. Daniel Fernández dreamed that an avalanche was coming down the mountainside and awoke with a start to find that what covered him was not snow but sheets and blankets. He tried to sleep again but felt uncomfortable. What damned fool is sticking his feet into my body? he thought to himself, and then once again awoke with a start to find himself alone in his hospital bed.

Coche Inciarte slept more soundly than Fernández

and was awakened by the sound of singing birds. He lay back, amazed and happy, and when a nurse came into his room he asked her to open the window. She did so, and he breathed the fresh air into his lungs. At the same time the survivors who were healthier than he was came out of their rooms and sat on the wicker chairs at the end of the corridor, where the windows looked out onto the foothills of the Andes.

At eight o'clock Father Andrés returned to the hospital with a cassette tape recorder with which he recorded statements from the survivors. "We had an enormous desire to survive," said Mangino, "and faith in God. Our group was always united. When the spirits of one went down, the rest made sure to raise them. Praying the rosary every night strengthened the faith of all of us, and this faith helped us get through. God gave us this experience to change us. I changed. I know now that I shall be different to what I was . . . all thanks to God."

"We hope to preach faith to the world," said Carlitos Páez. "Although this experience was sad because of all the friends we lost, it has helped us a lot—in fact it is the greatest experience of my whole life. As far as the trip is concerned, I shall never go on a plane again. I'm going back by train. . . . I have had a lot of experience as a rugby player. When you make a try, it isn't you but the whole team that has scored. That's the best thing about it. If we were able to survive, it was because we all acted with team spirit, with great faith in God—and we prayed."

At half past ten a press conference was held on a terrace outside the private wing for the horde of desperate journalists who had been besieging the hospital since the day before. Inciarte and Mangino remained in their beds, but the other survivors allowed themselves to be photographed, for they were now dressed in clothes that the hospital had bought from—or, more often, been given by—the merchants of San Fernando. The conference was short and little was said. When they were asked what they had eaten to stay alive, they answered that they had bought a lot of cheese in Mendoza and that herbs grew in the mountains.

At eleven o'clock, mass was celebrated by the Bishop of Rancagua and three other priests in the brick church adjoining the hospital. The survivors, some in wheelchairs, were in the front row of the congregation. It was a momentous occasion for them all, and their emaciated faces were stretched with expressions of the love and gratitude that they felt toward God. In all the weeks that they had waited for this day, never for a moment had they lost their faith in Him; never had they doubted His love or His approval of their desperate and ugly struggle to survive. Now those same mouths which had eaten the bodies of their friends hungered for the body and blood of Christ; and once again, from the hands of the priests of their church, they received the sacrament of Holy Communion.

After the ceremony they prepared to leave for Santiago, for by then it had been decided that, while Mangino and Inciarte would be taken by ambulance to the Posta Central, the other six could go straight to the Sheraton San Cristóbal Hotel, where all the Uruguayans were to celebrate Christmas.

Before departing, some of the survivors accepted invitations to lunch from various citizens of San Fernando. The Canessa family went to the home of Dr. Ausin, while Parrado, along with his father, sister, and brother-in-law, went to a restaurant with a Mr. Hughes and his son Ricky, after which they were driven the ninety-odd miles to Santiago in a Chevrolet Camaro—a joy for Parrado which was in no way diminished by his experiences in the Andes.

Javier Methol was the first of the second group of the survivors to enter the Posta Central in Santiago. A ward had been found for them on the top floor of the hospital, and since the helicopter landed on the roof he had only to be carried down one flight of stairs. All the same the wide corridors were crammed with people smiling, applauding, even weeping for joy at the sight of the young Uruguayans who had been so miraculously restored to life.

No sooner had he been shown his bed than Methol—still wearing the same clothes he had worn in the Andes—asked if he could take a shower.

"Of course," said the nurse, and she took him in a wheelchair to a nearby bathroom. She then explained that since she was responsible for him she would have to stay with him while he took his shower. Methol set her at ease. Had he been the most modest man in the world, a row of nurses would not have stopped him from taking a shower. He stripped off his filthy clothes and stepped under the strong jets of hot water. They whipped the skin off his emaciated back and shoulders, but it was pain which was a joy to bear. When he emerged from the shower and put on a white hospital smock, he felt like a new man. He sat once again in the wheelchair and was pushed by his nurse back into the ward, where he saw a group of his fellow survivors still wearing their old clothes.

"Oh, please," said Methol, "please take these dirty people out of here."

As soon as their patients were washed, the doctors of the Posta Central examined them, taking X rays and blood tests. Their verdict was that everyone except Harley and Methol could be allowed to move to the Sheraton San Cristóbal Hotel that afternoon. These two were placed in a ward with Inciarte and Mangino, who had arrived from San Fernando. Of the four, Roy's condition was the most critical, for blood tests revealed a severe deficiency of potassium which endangered his heart.

The others, however, were not only well but almost pathologically high-spirited. Gustavo Zerbino escaped from the hospital to find some shoes, accompanied by his father, whom he met on the way out. Moncho Sabella drank a bottle of Coca-Cola which caused his stomach to swell up. He also suffered from the zeal of a young nurse who was so eager to do something for the young Uruguayans that she tried to take blood from Moncho's arm without seeming to know where to find the vein with the needle. Moncho put up with this scientific investigation—though his arm hurt for three days afterward—but he, along with his companions, was quite sure in his own mind what their medicine should be; while the doctors analyzed their condition, they asked for food.

The nurses responded with some tea, biscuits, and cheese. An immediate request was made for more cheese. It came, and shortly afterward they were served lunch—first steak with mashed potatoes, tomatoes and mayonnaise, then gelatin. The gelatin was eaten in a minute and the hungry patients asked for more. Then they asked for Christmas cake but were told that it was not to be had. Now they must rest.

At seven in the evening, after a mass in the amphitheater of the Posta, Delgado, Sabella, Francois, Vizintín, Zerbino, and Fito Strauch set out to join the others in the Sheraton San Cristóbal Hotel. At nine that evening, those who had remained were rewarded for their patience by a piece of Christmas cake. The nurses also told them that at eleven they would bring them a surprise; sure enough, at the appointed hour they were each given a most delicious chocolate mousse topped with cream—something the nurses themselves were to have eaten that night. The four ate the mousse, savoring each spoonful, and then went to sleep as happy men.

An hour later Javier Methol awoke. His stomach was in turmoil. He rang for a nurse and asked for something to settle his digestion. The nurse brought a potion which he swallowed, but an hour later he awoke again to find that he was afflicted by the most terrible diarrhea. He was paying the price for the mousse.

3

By the evening of December 23, the whole party of Uruguayans who had come to Chile upon hearing the news of the rescue had settled in Santiago—the survivors with their parents and relatives in the Sheraton

San Cristóbal Hotel on the edge of the city, the parents and relatives of those who did not survive in the more old-fashioned Crillon Hotel in the center.

There, at the Crillon, the father of Gustavo Nicolich opened the two letters given to him by Zerbino that his son had written on the mountain: "One thing which will seem incredible to you—it seems unbelievable to me—is that today we started to cut up the dead in order to eat them. There is nothing else to do." And then, a little later, the words with which the boy had so nobly predicted his own fate: "If the day came and I could save someone with my body, I would gladly do it." This was the first intimation that any parent at the Crillon had had that it was the bodies of their sons which had kept the sixteen alive, and Nicolich, already punch-drunk with grief at the death of his son, recoiled still further at the dread implications of the letter. Considering at that moment that the truth might never be known, he removed the sheet of the letter (which was addressed to Gustavo's *novia,* Rosina Machitelli) and concealed it.

Meanwhile, in the Sheraton San Cristóbal, the twelve survivors who had been let out of the hospital were basking in a plenitude of all that had for so long been denied them. Half of them were reunited with their parents. Pancho Delgado and Roberto Canessa were together once again with their loyal *novias,* Susana Sartori and Laura Surraco. In the Posta Central, Coche Inciarte was with Soledad González. The hotel itself was a total contrast to the Fairchild. It was a brand-new building, overlooking Santiago, with the smell and feel of utmost luxury. There was a swimming pool, and of course a restaurant, and it was of this latter facility that the twelve boys most immediately availed themselves. When Moncho Sabella arrived at the Sheraton on the afternoon of the twenty-third, he found Canessa already installed, eating a large plate of shrimps. Moncho immediately sat down with his brother, who had flown out from Montevideo, and ordered a bowl of shrimps for himself. Shortly after they had eaten them both boys were sick, but this did not spoil their appetite. They immediately ordered more food and started all over again with steaks, salads, cakes, and ice cream.

On the other hand, neither Sabella nor Canessa was overwhelmed by this luxury. When Dr. Surraco remarked to Canessa that the hotel must seem extraordinarily comfortable after the hulk of the Fairchild, Canessa retorted that it made no difference to him whether he was in a Sheraton hotel eating shrimps and ice cream or a shepherd's hut eating cheese.

The parents and relatives of the twelve boys were so happy to have them back among the living that they raised no objection to this compulsive indulgence of pathological greed. They were already aware that their sons and *novios* were not likely to behave as if they had just returned from a summer holiday. The long weeks of suffering and starvation had left their mark on the boys' behavior; like spoiled children some would tolerate no restraint, and when not indulging the more overt emotions of joy and delight at this reunion they would lapse into sharpness and irritability—above all with their parents, whose concern for their well-being annoyed them. Had they not proved that they could look after themselves?

These feelings were exacerbated by some parents' reaction to the anthropophagous aspect of the Christmas Miracle. Unprepared for the news, they had been shocked and for the most part never alluded to it again. They also quite clearly dreaded that the news would break on the outside world, and though some of the survivors conceded to themselves that their parents' reaction was only to be expected, they were all decidedly upset and injured that anyone should be appalled at what they had done. They read into those involuntary expressions of shock and disgust a preference for the alternative—which was that all of them should have died and none been eaten.

Their peace of mind was not assisted by the presence in the hotel of a mass of journalists and photographers asking incessant questions and taking pictures of them whenever they moved, ate, or kissed their parents. And more agonizing still were the equally persistent questions of the relatives of the boys who had not returned—the parents of Gustavo Nicolich and Rafael Echavarren; the brothers of Daniel Shaw, Alexis Hounie, and Guido Magri—who came up from the Crillon Hotel to discover the exact circumstances of the death of their brothers and children. It was not something which, at that moment, the survivors wished to remember and discuss.

Nor were they at all acclimated to the civilization of the Sheraton San Cristóbal Hotel. They were extremely uncomfortable in the large, soft beds, for they were used to sleeping in contorted positions. That night Sabella awoke every half hour and, finding himself awake, rang room service for some food. It was a hard night for him—and a hard night for his brother, who was sleeping in the same room.

The next day, December 24, the four who had remained in the Posta Central were released and joined the others in the Sheraton, but the sixteen were only reunited for a short time, for the Francois family and Daniel Fernández had decided to return at once to Montevideo. Though Daniel's two uncles and aunts were there in Santiago, he wanted to see his parents and thought it unnecessary and extravagant for them to come to Santiago. He therefore caught a scheduled KLM flight to Montevideo. The father and brother of Daniel Shaw were on the same plane.

A party of the other boys decided that they needed some more clothes and wanted to hire a taxi, but none of the Chileans would hear of it and insisted on taking them downtown in their own cars. There they walked in the streets, looking into the store windows. They were identifiable to all as the survivors, for being used to deep snow they walked ponderously, like penguins. Whenever they were recognized they were greeted with such joy and kindness by the citizens of Santiago that they might have been their own sons who had been saved from the Andes.

When they went into a clothing store and presented their purchases, the owners would not take money but insisted that they accept the garments as gifts. It was the same when they came to a long line which because of a shortage had formed outside a cigarette stand; an old man at the head of the line insisted that they accept his pack of cigarettes.

Again, when they returned to the hotel (Parrado

walked back from the center of Santiago) and a group of them sat down to lunch and ordered a bottle of white wine, the Chileans sitting at the next table took their own bottle and gave it to them. At the bar they would be plied with whiskey and champagne—and in the lobby of the hotel a little boy presented them with a large box of chewing gum.

They were admired and feted not just as heroes who had endured and triumphed over the awesome Andes which loom forbiddingly along the entire length of Chile but as the living embodiment of an apparent miracle. The cheese and the herbs upon which they said they had lived seemed as paltry a source of nourishment as the loaves and fishes in the Gospel. Their survival seemed incontrovertibly miraculous. One woman whose son was sick even came to the hotel in the belief that if she could only embrace one of the survivors her son would be cured.

That evening the Christmas party organized by César Charlone was held. It was a moment of intense emotion for them all. Only four days before there had seemed to be no hope that the parents would ever see their children again, or the children spend that Christmas with their parents. Now they were together. The burning faith of Madelón Rodríguez, Rosina Strauch, Mecha Canessa, and Sarah Strauch; the heroic searching of Carlos Páez Vilaró, Jorge Zerbino, Roy Harley, and Juan Carlos Canessa—all now had their reward in the living hands and lips and bodies of their sons. As with Abraham and Isaac, God had excused them the sacrifice of their sons at the very moment when the Christian world prepared to celebrate the birth of His own.

Later that night a Uruguayan Jesuit who taught theology at the Catholic University in Santiago came to the hotel at the invitation of Señora Charlone to talk to some of the survivors, in preparation for the mass that was to be held for them the next day. As it was, Father Rodríguez remained talking with Fito Strauch and Gustavo Zerbino until five in the morning. They told him that they had eaten the bodies of their friends to stay alive and, like Father Andrés in San Fernando, Father Rodríguez did not hesitate to endorse the decision they had made. Whatever doubts there might have been about the morality of what they had done were dissipated in his mind by the sober and religious spirit in which they had made their decision. The two boys told him what Algorta had said when they had cut meat from the first body, and while the Jesuit discounted any strict correlation between cannibalism and communion, he was moved as so many others had been by the pious spirit which was manifest in the dictum.

The Christmas mass was held at the Catholic University at twelve the next day, and the sermon delivered by Father Rodríguez, though it made no mention of anthropophagy, was an unequivocal affirmation of what the young men had done to stay alive. Though not all the boys or their parents were acquainted with Karl Jaspers, or the concept of a limited situation, they all believed in the authority of the Catholic Church and were profoundly reassured by what was said.

It was the calm before the storm. The continuing celebration of Christmas after the mass was over marked the last untroubled hours they were to spend in Santiago. Journalists from all over the world continued to hover around them like the condors in the Andes, and it was quite clear to the Uruguayans that they had not yet caught the scent of their real prey. It was not that the boys or their parents conspired—beyond their lame lies about the herbs and cheese—to conceal what they had done; it was just that they hoped the news could be kept until they were back in Montevideo.

The story broke in a Peruvian newspaper and was immediately picked up by the Argentinian, Chilean, and Brazilian papers. As soon as the journalists in Santiago sniffed the story, they fell once again upon the survivors and asked if it was true. Confused, the boys continued to deny it, but those who had betrayed their secret had furnished the proof, and on December 26 the Santiago newspaper *El Mercurio* published on its front page a photograph of a half-eaten human leg lying in the snow against the side of the Fairchild. The boys conferred as to what they should do and decided that, rather than talk about what had happened to any particular newspaper, they would hold a news conference when they returned to Montevideo. Since they had been in touch with the president of the Old Christians, Daniel Juan, they agreed that the conference should be held at their old school, the Stella Maris College.

These were frail defenses against the tornado that raged around them. The news—which had been given to the papers by the Andinists—merely whetted the appetites of the world's press, and the boys in the hotel were bombarded with questions which they would not answer. Indeed, they became increasingly disgusted with the journalists, who showed no reticence or tact in what they asked. There were even persistent suggestions by an Argentine journalist that the avalanche had not occurred but had been invented to conceal the fact that the stronger boys had killed the weaker ones to provide themselves with food.

The survivors were still exceedingly vulnerable, and these assaults upset them. Moreover, they saw that a Chilean magazine which usually specialized in pornography had taken two pages to print photographs of limbs and bones which had lain around the Fairchild. Another Chilean newspaper printed the story under the headline: "May God Forgive Them." When some of the parents saw this, they wept.

The atmosphere in the Sheraton San Cristóbal was poisoned by this clamor. The survivors were impatient to return to Montevideo and reluctantly agreed to fly rather than go by bus and train. Charlone (who had never been forgiven by some of the parents for what they considered to be poor treatment of Madelón Rodríguez and Estela Pérez) arranged for a Boeing 727 of LAN Chile to take them on December 28. Before that, however, Algorta left with his parents to stay with friends outside Santiago. Parrado too left the Sheraton San Cristóbal with Juan and Graciela and his father—first to the Sheraton Carrera in the center of Santiago, then to a house in Viña del Mar which had been lent to them by friends. He was tired of being photographed every moment of the day and disgusted at the journalists' callous questions. Even the incessant celebrations were somehow depressing, for though he was alive, the two women whom they had all loved most in the world remained as frozen corpses in the Andes.

Fifteen

1

The story of the survival of the young Uruguayans, after ten weeks in the Andes, had been sensational enough to interest the newspapers and radio and television stations of the whole world, but when the news broke that their survival had depended on eating the dead these same media went wild. The story was broadcast and printed in almost every nation in the world, with one notable exception—Uruguay itself.

There had been reports, of course, of the discovery and rescue of the survivors, but when rumors of cannibalism reached the news desks of the papers in Montevideo they were treated first with skepticism and then with reticence. There was at that time no censorship of the press (beyond a ban on any mention of the Tupamaros); the decision by the Uruguayan journalists to wait until their fellow countrymen had returned to Montevideo and given their account of what had happened can only be explained as the product of a spontaneous patriotic reserve.

This is not to say that there were not journalists eager to discover whether the rumors were true, but since most of the survivors were still in Santiago it was not an easy thing to do. Daniel Fernández, however, was already in Montevideo. He had been met at the airport by his parents, who had driven him to their flat and refused to allow any visitors. By the next day, however, the whole block was besieged by friends and journalists eager to see him. It was Christmas Day and the Fernándezes could not keep their door closed forever, so they opened it to admit one friend—but once open it could not be shut again. A horde of journalists and acquaintances poured into the apartment, and Daniel agreed to be interviewed.

He sat facing the group of journalists, and one of them suddenly handed him a piece of paper and asked him to read it. Daniel unfolded it and saw a Telex message with the news that he and the other fifteen survivors had eaten human flesh.

"I have nothing to say about that," he said.

"Can you confirm or deny it?" asked the journalist.

"I have nothing to say until my friends are back in Uruguay," said Daniel.

While this exchange was taking place, Juan Manuel Fernández read the Telex. "The man who wrote this is a son of a bitch and the man who brought it here is even more of a son of a bitch," he said, most forcefully. He was about to show the journalist unceremoniously to the door, but a friend of Daniel's restrained him and the journalist departed of his own accord.

After he had left, Fernández took his son aside and said, "Look, now, you must say that this isn't true."

"It is true," said Daniel.

The father looked abruptly at the son with an expression of mild distaste on his face; but later, when he realized that it was something his son had done from necessity, he got used to the idea and was surprised that it had not occurred to him before.

2

The Boeing 727 of LAN Chile which had been chartered by Charlone to fly the survivors and their families back to Montevideo was given the elite crew used when President Allende himself went abroad. It stood ready on the tarmac of Pudahuel airport on the afternoon of December 28 while its sixty-eight passengers were given an emotional and ceremonious farewell by the Chileans, who had on the whole treated them so well.

They boarded the plane at four o'clock but were forced to wait an hour before takeoff. The first reason for delay was Vizintín, who had been kept in Santiago by an interview; then there were the weather reports from the cordillera. These were still unfavorable, but rather than alarm the survivors the crew told them that they had run out of fruit juice and had to replenish their supplies.

Vizintín arrived, but still the plane stayed on the tarmac. The survivors were nervous and tense as they strapped themselves into their seats. Most of them had wanted to return overland and had only consented to go by plane because the journey through the Andes and across Argentina by train was considered dangerous in their present state of health.

At last the weather reports were favorable and the

plane took off. A short time later the pilot, Commander Larson, announced that they were over Curicó, but no one accepted his invitation to come to his cabin and look down on the town whose name had meant so much to them. As a group they were nervous not just because they were up in an airplane again but because they were uncertain of what lay ahead in Uruguay. They talked compulsively among themselves and to the two Chilean journalists who were traveling with them.

One of these—Pablo Honorato from *El Mercurio*—sat next to Pancho Delgado, who, when the plane began to land at Carrasco airport, became even more nervous than he had been and grabbed Honorato. But then there arose shouts of "*¡Viva Uruguay!*" and then "*¡Viva Chile!*" to keep up the courage of the survivors. As the plane circled over Montevideo, and they saw once again the muddy waters of the River Plate and the roofs and streets of their beloved city, they began to sing their national anthem:

Orientals, our country or the grave,
Liberty, or death with glory. . . .

As the last word burst from their lips, the plane touched down on Uruguayan soil.

The plane taxied across the tarmac and came to a stop outside the same airport building that they had left so optimistically almost eleven weeks before. The differences between that departure and this return were many; while only one or two members of their families had come to see them off, the whole city of Montevideo seemed now to be there to greet them, including the wife of the President of Uruguay. The balconies on the airport building were lined with shouting, waving people, and there were lines of police to keep this crowd from surging onto the tarmac.

The survivors and their families were ushered into buses which drove up alongside the airplane. The boys wanted to be driven in front of the balconies so that they could salute their friends, but on the instructions of the Army the buses drove straight out of the airport toward the Stella Maris College.

Everything was ready for their arrival. The large brick assembly hall, designed by the father of Marcelo Pérez, had been laid out as for a prize-giving, with a long table on a podium and a system of microphones and loudspeakers which would enable the many journalists who were already seated facing this stage to hear what was said. It was not only the Christian Brothers who had done this but also the officers of the Old Christians, who greeted the survivors as the coaches turned into the driveway of the school.

It was an emotional reunion and one in which the turmoil of the situation—with cameras whirling and clicking all around them—could not blunt the grievous truth that among those who now climbed off the bus and took their places on the podium there were only three members of the rugby team which had set out for Chile: Canessa, Zerbino, and Vizintín. Parrado and Harley were still in Chile. As Daniel Juan and Adolfo Gelsi looked at the thin, bearded faces, they searched for their champions—Pérez, Platero, Nicolich, Hounie, Maspons, Abal, Magri, Costemalle, Martínez-Lamas, Nogueira, Shaw—but they were not there.

Nevertheless the whole group of survivors had entrusted the press conference to the care of the Old Christians, and with calm mastery of a potentially chaotic situation—a room packed with journalists from all over the world, parents, parents of the dead, friends, relations, and television cameras—Daniel Juan took his seat on the center of the podium and the conference began.

The survivors had decided that they would speak in turn, each of them taking a particular aspect of their experience, and when they had finished they would ask the Uruguayan press if they wished to question them further. The only dispute among them was as to how they should treat the question of cannibalism. Some of the boys and their parents thought that they should be quite frank about what had happened; others considered that it would be enough for some vague allusion to be made to it. A third group—notably Canessa and his father—thought that no mention should be made of it at all.

A compromise was reached: Inciarte would speak about it. He offered to do so, and it was agreed that he was the most suitable person because of his high-minded attitude to what had happened, but on the day of the conference itself Coche began to have doubts about his own abilities. He stuttered, and he was afraid that in front of all the journalists and cameras he would break down. Pancho Delgado volunteered to take his place.

The conference began. The whole room listened in silence as, one after the other, the survivors told their heroic and tragic story, until it was Delgado's turn. Almost at once his eloquence—which had been of such little use on the mountain—came into its own.

"When one awakes in the morning amid the silence of the mountains and sees all around the snowcapped peaks—it is majestic, sensational, something frightening—one feels alone, alone, alone in the world but for the presence of God. For I can assure you that God is there. We all felt it, inside ourselves, and not because we were the kind of pious youths who are always praying all day long, even though we had a religious education. Not at all. But there one feels the presence of God. One feels, above all, what is called the hand of God, and allows oneself to be guided by it. . . . And when the moment came when we did not have any more food, or anything of that kind, we thought to ourselves that if Jesus at His last supper had shared His flesh and blood with His apostles, then it was a sign to us that we should do the same—take the flesh and blood as an intimate communion between us all. It was this that helped us to survive, and now we do not want this—which for us was something intimate, intimate—to be hackneyed or touched or anything like that. In a foreign country we tried to approach the subject in as elevated a spirit as possible, and now we tell it to you, our fellow countrymen, exactly as it was. . . ."

As Delgado finished, it was quite evident that the entire company was deeply moved by what he had said, and when Daniel Juan asked the assembled journalists if they had any questions to ask the survivors he was told that there were none. Whereupon the whole room burst into a spontaneous hurrah for the gentlemen of the Uruguayan and international press, followed by a final cheer for those who had not returned.

3

With the conclusion of the press conference, the public ordeal of the survivors, which had followed so closely on their private ordeal, came to an end, and they were able at last to return to the homes and families of which they had dreamed while imprisoned high up in the Andes.

It was not easy to adapt to the reality. Their experience had been long and terrible; its effect had gone deep into both their conscious and unconscious minds and their behavior reflected this shock. Many of the boys were brusque and irritable with their parents, *novias*, and brothers and sisters. They would flare up at the least frustration of their smallest whim. They were often moody and silent or would talk compulsively about the accident. Above all, they would eat. No sooner was a dish set on the table than they would attack it, and when a meal was over they would stuff themselves with sweets and chocolates so that Canessa, for example, became bloated in the space of only a few weeks.

Their parents felt helpless in the face of this behavior. Some had been warned by the psychiatrists in Santiago who had briefly examined some of their sons that they might face some difficulty in readapting to normal life and that there was little they could do to help. Their case, of course, was as baffling to psychiatrists as it was to the parents themselves, for there were few case histories relating to a breach of this particular taboo. No one could know what the effect would be on their minds; they could only wait and see.

Some of the parents were also in a state of shock. It was as if they were paralyzed by surprise and gratitude at the sight of those sons whom they had given up for dead. Coche Inciarte's mother, for example, was unable to take her eyes off her son as he ate. At night she lay in the same room as he did but did not close her eyes; she simply watched her son as he slept.

The mothers who were best equipped to deal with the unique situation in which they found themselves were Rosina and Sarah Strauch and Madelón Rodríguez. Not only did these three women have strong personalities which nothing could intimidate; they also looked upon the whole saga as theirs as well as their sons'. They behaved, as it were, as if their faith and their prayers were as much responsible for the survival of the boys as the boys' own efforts. They were quite decided in their own minds on something which still confused the boys themselves—the meaning of what they had been through. To these three the boys had disappeared and then returned to prove to the world the miraculous powers of the Virgin Mary—in the case of the Strauch sisters, the Virgin of Garabandal.

The beneficiaries of this miracle were justifiably confused because other interpretations were put forward. On the one hand they were aware that many—especially among older people—were appalled by what they had done and considered that they should have chosen to die. Even Madelón's mother, who as much as anyone had believed in the return of her grandson, could not bring herself to contemplate this aspect of his survival.

The Catholic Church, however, was quick to dismiss this primitive reaction. "You cannot condemn what they did," said Monsignor Andrés Rubio, Auxiliary Bishop of Montevideo, "when it was the only possibility of survival. . . . Eating someone who has died in order to survive is incorporating their substance, and it is quite possible to compare this with a graft. Flesh survives when assimilated by someone in extreme need, just as it does when an eye or heart of a dead man is grafted onto a living man. . . . What would we have done in a similar situation? . . . What would you say to someone if he revealed in confession a secret like that? Only one thing: not to be tormented by it . . . not to blame himself for something he would not blame in someone else and which no one blames in him."

Carlos Partelli, the Archbishop of Montevideo, confirmed his opinion. "Morally I see no objection, since it was a question of survival. It is always necessary to eat whatever is at hand, in spite of the repugnance it may evoke."

And finally the theologian of *L'Osservatóre Romano*, Gino Concetti, wrote that he who has received from the community has also the duty to give to the community or its individual members when they are in extreme need of help to survive. Such an imperative extends especially to the body, which is otherwise consigned to dissolution, to uselessness. "Considering these facts," Father Concetti went on, "we justify on an ethical basis the fact that the survivors of the crash of the Uruguayan airplane fed themselves with the only food available to avoid a sure death. It is legitimate to resort to lifeless human bodies in order to survive."

On the other hand, the church did not concur with the view that had been expressed by Delgado at the press conference that eating the flesh of their friends was equivalent to Holy Communion. When Monsignor Rubio was asked whether a refusal to eat the flesh of a dead human being could be interpreted as a form of suicide, and the opposite as an act of communion, he replied, "In no way can it be understood as suicide, but the use of the term communion is not correct either. At most it is possible to say that it is correct to use this term as a source of inspiration. But it is not communion."

It was clear, therefore, that the survivors were to be regarded neither as saints nor as sinners, but a role was increasingly sought for them as national heroes. The newspapers and radio and television stations began to take an understandable pride in what these young fellow countrymen had achieved. Uruguay was a small nation in a large world, and never since their soccer victory in the World Cup in 1950 had the activities of any Uruguayans achieved such world renown. There were many articles describing their courage, endurance, and resourcefulness. The survivors, on the whole, rose to the occasion. Many of them kept their beards and long hair and were not ill pleased to be recognized wherever they went in Montevideo and Punta del Este.

Though every interview and article emphasized that their achievement had been the work of the whole group, it was inevitable that some of the survivors should fit the role of national hero more successfully than others. Some, for example, were not even on stage. Pedro Algorta had gone to join his parents in Argentina. Daniel Fernández had retired to his parents' *estancia* in the country. His two cousins, Fito and Eduardo Strauch, were too taciturn to project for the public an image which corresponded to the part they had played on the mountain.

The ablest exponent of the whole experience was Pancho Delgado, and it was quite natural, because he was the one who had dealt with the question of cannibalism at the press conference, that the press should look to him for further information. Delgado rose to the occasion. He went by bus to Río de Janeiro (with Ponce de León) to appear on television and gave extensive interviews to the Chilean magazine *Chile Hoy* and the Argentinian review *Gente*. It was not surprising either that Delgado, finding himself once again in a situation where his talents could bc useful, should use them; nor that the press should take advantage of so eloquent a survivor. His prominence in the public eye, however, did not endear him to his former companions.

The other member of the group whose behavior some felt was unseemly was Parrado. His character had undergone a greater metamorphosis than that of the others. The timid, uncertain boy had emerged from the ordeal as a dominating, self-assured man who was everywhere recognized and acclaimed as the hero of the Andean odyssey, but the man still contained the tastes and enthusiasms of the boy and, being freed now from his close acquaintance with death, he was determined to indulge them.

Thinking him dead, his father had sold his Suzuki

500 motorcycle to a friend, but so pleased was he to see Nando return from the grave that he bought him an Alfa Romeo 1750. In it Parrado roared off along the coast to Punta del Este to lead the sweet life on the beaches and in the cafés and nightclubs of the glamorous resort. All the beautiful girls who had previously thought of him as the shy friend of Panchito Abal now flocked around him and vied with one another for his attention. Parrado did not hold back. The only thing he permitted to draw him away from Punta del Este was the Formula One races in Buenos Aires. There he met drivers Emerson Fittipaldi and Jackie Stewart, and they were photographed together, for everywhere Parrado went he was followed by a horde of journalists and photographers.

These pictures all appeared in the Uruguayan papers and dismayed his fifteen companions. When the paper showed him among a gaggle of bathing beauties in Punta del Este as a judge of a beauty competition, they voiced their objection and Parrado withdrew; to him, as to the others, the unity of the sixteen was still of the greatest importance.

While he recognized that it was their combined efforts which had saved their lives, however, Parrado felt that his own achievement was a triumph which he should be allowed to celebrate. Life had conquered over death and should be lived to the full in the same way as he had lived it before . . . but of course some things had changed forever. One evening in the middle of January he went into a nightclub with a friend and two girls. It was a place he had frequented with Panchito Abal, and he had not been there since the crash. As he sat down at the table and ordered drinks, he was suddenly struck with the truth that Panchito was dead, and for the very first time in the three months of trial and suffering, he burst into tears. He fell forward onto the table and cried and cried and cried. He could not stem the flood of tears, so the four of them left the nightclub. Soon after that, Parrado started work again, selling nuts and bolts at La Casa del Tornillo.

The reason why the other fifteen survivors looked askance at Parrado's return to the kind of life he had led before was that they themselves had a more elevated—almost a mystical—concept of their experience. Inciarte, Mangino, and Methol felt certain that they were the beneficiaries of a miracle. Delgado considered that to have lived through the accident, the avalanche, and the weeks which followed could be ascribed to the hand of God, but that the expedition was more a manifestation of human courage. Canessa, Zerbino, Páez, Sabella, and Harley all felt that God had played a fundamental role in their survival; that He had been there, present, on the mountain. On the other hand, Fernández, Fito and Eduardo Strauch, and Vizintín were more inclined to believe in all modesty that their survival and escape could be ascribed to their own efforts. Certainly, prayer had assisted them—it had been a bond which held them together and a safeguard against despair—but if they had relied on prayer alone they would still be up on the mountain. Perhaps the greatest value of the grace of God had been to preserve their sanity.

The two most skeptical about the role God had played in their rescue were Parrado himself and Pedro

Algorta. Parrado had good reason, for like many of them he could see no human logic in the selection of the living from the dead. If God had helped them to live, then He had allowed the others to die; and if God was good, how could He possibly have permitted his mother to die, and Panchito and Susana to suffer so terribly before their death? Perhaps God had wanted them in heaven, but how could his mother and sister be happy there while he and his father continued to suffer on earth?

Algorta's case was more complex, for his Jesuit education in Santiago and Buenos Aires had left him better equipped to deal with the mysteries of the Catholic faith than had the simpler theological education of the Christian Brothers. Moreover he had been, before he left, among the more earnestly religious of the passengers on the plane. He had not had the easy though somewhat unorthodox familiarity with God of Carlitos Páez, but the orientation of his life—especially his political convictions—was centered around the precept that God is love. After seventy days in the wilderness of the Andes he did not believe any the less that God was love, but it had taught him that the love of God was not something to count on for survival. No angels had come down from heaven to help them. It was their own qualities of courage and endurance which had seen them through. If anything, the experience had made him less religious; he now had a stronger belief in man.

They all agreed, however, that their ordeal on the mountain had changed their attitude toward life. Suffering and privation had taught them how frivolous their lives had been. Money had become meaningless.

No one up there would have sold one cigarette for the five thousand dollars which they had amassed in the suitcase. Each day that passed had peeled off layer upon layer of superficiality until they were left only with what they truly cared for: their families, their *novias,* their faith in God and their homeland. They now despised the world of fashionable clothes, nightclubs, flirtatious girls, and idle living. They determined to take their work more seriously, to be more devout in their religious observances, and to dedicate more time to their families.

Nor did they intend to keep what they had learned to themselves. Many of them—especially Canessa, Páez, Sabella, Inciarte, Mangino, and Delgado—felt a sense of vocation to make use of their experience in some way. They felt touched by God and inspired by Him to teach others the lesson of love and self-sacrifice which their suffering had taught them. If the world had been shocked by the knowledge that they had eaten the bodies of their friends, this shock should be used to show the world just what it can mean to love one's neighbor as oneself.

There was only one rival, as it were, for the lesson which was to be drawn from the return of the sixteen survivors, and this was the Virgin of Garabandal, for whatever the opinions of their sons there remained that group of strong-minded women who had invoked her intercession and now felt that she had answered their prayers. They remembered when the skeptics had conceded that only a miracle could save the boys, and they were determined not to see their Virgin cheated of it just because it was susceptible to a rational explanation of a somewhat disagreeable nature. Indeed, they gripped the nettle of cannibalism with the thesis that the manna from heaven which had rained down upon the deserts of Sinai was but a euphemistic description for God's inspiration to the Jews to eat the bodies of their dead.

4

Twenty-nine of those who had left in the Fairchild had not returned, and for the families of those twenty-nine the return of the sixteen meant the confirmation of their death. It was, moreover, a confirmation of a disturbing nature. The Abals learned of the physical suffering of their son; the Nogueiras faced the mental agony of theirs. Every member of every family confronted the knowledge that their husbands, mothers, and sons were not only dead but might have been eaten.

It was a bitter admixture to hearts already brimful with sorrow, for however noble and rational the mind may have been in contemplation of this end, there was a primitive and irrepressible horror at the idea that the body of their beloved should have been used in this way. For the most part, however, they mastered this repugnance. The parents showed the same selflessness and courage as their sons had done and rallied around the sixteen survivors. Dr. Valeta, the father of Carlos, went with his family to the press conference and afterward spoke to the newspaper *El Pais*. "I came here with my family," he said, "because we wanted to see all those who were the friends of my son and because we are sincerely happy to have them back among us. We are glad, what is more, that there were forty-five of them, because this helped at least sixteen to return. I'd like to say, furthermore, that I knew from the very first moment what has been confirmed today. As a doctor I understood at once that no one could have survived in such a place and under such conditions without resort to courageous decisions. Now that I have confirmation of what has happened I repeat: Thank God that the forty-five were there, for sixteen homes have regained their children."

The father of Arturo Nogueira wrote a letter to the papers:

Dear Sirs:

These few words, written in obedience to what is in our hearts, want to pay tribute, with homage, admiration, and recognition, to the sixteen heroes who survived the tragedy of the Andes. Admiration, because this is what we feel before the many proofs of solidarity, faith, courage, and serenity which they had to face and which they overcame. Recognition, profound and sincere, because of the care they gave in every moment to our dear son and brother Arturo up to the time of his death many days after the accident. We invite every citizen of our country to spend some minutes in meditation on the immense lesson of solidarity, courage, and discipline which has been left to us by these boys in the hope that it will serve us all to overcome our mean egotism and petty ambitions, and lack of interest for our brothers.

The mothers showed similar courage. Some saw their dead sons in the survivors, for it was not difficult to understand that if their children had stayed alive and the others had died, the same thing would have happened; and that if all forty-five had survived the accident and avalanche, all forty-five would now be dead. They could imagine, too, the mental and physical anguish suffered by the survivors. All they wished now was that they should forget what they had been through. After all, it was not the sons or brothers or parents of their friends that they had eaten to survive. They had been already in heaven.

Most of the parents had resigned themselves to the death of their sons soon after the accident. There were some, however, who felt particularly cheated by fate. Estela Pérez had believed quite as firmly as Madelón Rodríguez and Sarah and Rosina Strauch that Marcelo was alive, yet while their faith had been rewarded, hers had not. It was also a mean and bitter twist of fate that Señora Costemalle, whose other son had drowned off the coast of Carrasco and whose husband had died suddenly in Paraguay, should now have lost the last surviving member of her family.

The parents of Gustavo Nicolich were tormented by the knowledge that their son had lived for two weeks after the accident. They also felt some animosity toward Gerard Croiset, Jr., who, they concluded, had sent them off on a false trail at a time when to continue toward the Tinguiririca and Sosneado mountains might have saved their son's life.

It was certainly true that the interpretation that had been put on Croiset's clairvoyance had misled the parents, but there were many things in what he had said which turned out to be true. He had seen some difficulty with officials over one of the boys' papers at Carrasco airport; there had been such an incident. He had said that the pilot was not flying the plane, and it was true that Lagurara, not Ferradas, was at the controls. The plane, he had said, lay like a worm with a crushed nose but no wings and the front door was half open. All this was true. Croiset had also accurately described the maneuvers which would be necessary for a pilot to see the wreck from the air, and he had said that the plane was near a sign reading danger and not far from a village with white Mexican-style houses. Though nothing of this sort was encountered by Parrado and Canessa on their walk into Chile, a later expedition, from Argentina, to the site of the accident found in the vicinity a sign reading danger and a small village, Minas de Sominar, with white Mexican houses.

The landscape around the aircraft as described by Croiset—the three mountains, one without a top, and the lake—was found by the parents, but forty-one miles *south* of Planchon, whereas the Fairchild had crashed forty-one miles *north* of Planchon. The plane was not under a mountain, nor in or near a lake, nor had the pilot flown toward a lake to make a forced landing. The accident was not due to a blocked carburetor, as Croiset had said, nor was the pilot alone in the cabin, and whether or not he had indigestion could not be known. There were other details Croiset had given, when under pressure from the parents to supply them with more information, that seemed in retrospect to have little relevance to the tragedy, but at least in giving them he had saved some of the mothers from despair.

The dreams of Señora Valeta also had been extraordinarily accurate, but the only extrasensory perception which events showed to be completely correct was that of the old water diviner whom Madelón's mother and Juan José Methol had visited in the impoverished Maroñas district of Montevideo. He had pointed on a map to a spot nineteen miles from the spa of Termas del Flaco, which was exactly where the Fairchild was found to have come down. Remembering this, Juan José Methol went to find the old man and rewarded him with gifts of meat and money, which he in his turn shared with his impoverished neighbors.

5

An investigation into the causes of the accident was conducted by the air forces of both Uruguay and Chile. Both blamed the crash on the human error of the pilot, who had begun his descent toward Santiago when still in the middle of the Andes. The actual spot where the plane had crashed was nowhere near Curicó. The mountain on which the boys had spent so many days lay on the Argentine side of the frontier, between the Cerro Sosneado and the Tinguiririca volcano. The fuselage had lain at about 11,500 feet; the mountain climbed by the expeditionaries was around 13,500 feet high. It was estimated that if the expeditionaries had followed the valley beyond the tail instead of climbing the mountain to the west, they would have come to a road in about three days (though the road which Canessa thought he saw as he climbed the mountain was almost certainly a geological fault). Only five miles to the east of the Fairchild there was a hotel which, though open only in the summer, was stocked with supplies of canned food.

The attempt to call help with the plane's radio, which in all had cost them more than two weeks on the mountain, could never have succeeded. The transmitter required 115 volts AC, normally supplied by an inverter. The current supplied by the batteries was 24 volts DC.

There was little in the way of a postmortem. Though some of the parents felt anger toward the Uruguayan Air Force for the incompetence of its pilots, it was not a moment in the political history of Uruguay to take on a branch of the armed forces. On the whole they accepted what had happened as the will of God and were grateful to Him for those who had returned, accepting the elevated view of what had happened which emanated from the survivors themselves.

Javier Methol, now that he was living at sea level, lost the dopiness which had afflicted him at the high altitude of the mountain. Like his former companions, he too believed that God had permitted them to survive for some purpose, and the first task he undertook was to make up to his children, insofar as he could, for the loss of their mother. He went to live with Liliana's mother and father, who had, as he had known they would, taken care of his children. Reunited with them, Javier was almost content, for though he continued to miss Liliana he knew that she was happy in heaven.

One evening in Punta del Este, he went walking along the beach with his three-year-old daughter, Marie Noel. She was skipping along at his side, chattering all the time, when suddenly she stopped and looked up at him.

"Papa," she said. "You came back from heaven, didn't you? But when is Mama coming back?"

Javier crouched down to the level of his little daughter and said to her, "You must try and understand, Marie Noel, that Mama is so nice, so very nice, that God needs her in heaven. She is so important that now she is living with God."

6

On January 18, 1973, ten members of the Andean Rescue Corps, together with Freddy Bernales of the SAR, Lieutenant Enrique Crosa of the Uruguayan Air Force, and a Catholic priest, Father Ivan Caviedes, were flown in helicopters to the wreck of the Fairchild. There they pitched camp, intending to spend some days on the mountain, and set about gathering up the remains of the dead. They climbed to the top of the mountain to recover those bodies which had remained there and had now been uncovered by the melting of the snow.

A spot was found about half a mile from the site of the accident which was sheltered from possible avalanches and had enough earth to make a grave. Here they buried those bodies which were still intact and all the remains of those which were not. A rough stone altar was built beside the grave, and over it was placed an iron cross about three feet high. The cross was painted orange and on one side of it in black was the inscription "The World to Its Uruguayan Brothers," while on the other side were painted the words "Nearer, O God, to Thee."

After saying mass, Father Caviedes made an address to the men who had assisted at the ceremony. Then the Andinists returned to the hulk of the Fairchild, splashed it with gasoline, and set it on fire. The plane burned quickly in the strong wind, and as soon as they were sure that it was properly alight the Chileans prepared to leave. The silence of the mountains had been broken all too frequently by the rumble of avalanches, and they judged it too dangerous to stay.

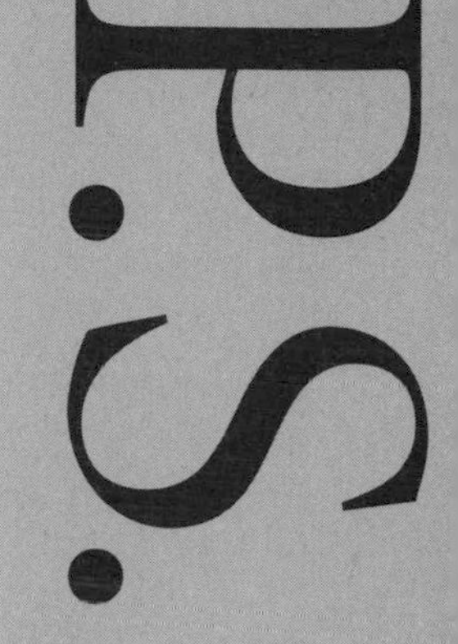

Insights,
Interviews
& More . . .

About the author

About the book

Read on

About the author

Meet Piers Paul Read

Piers Paul Read is the author of numerous critically acclaimed works of fiction and nonfiction, including *Ablaze: The Story of the Heroes and Victims of Chernobyl* (1993); *A Patriot in Berlin* (1996); *On the Third Day* (1989); *A Season in the West* (1989); and *The Templars: The Dramatic Story of the Knights Templar, the Most Powerful Military Order of the Crusades.*

A Discussion with Piers Paul Read
Three Decades Later

Although you touch on your relationship with the survivors during and soon after the writing of Alive, *have you maintained a relationship with any or all of the survivors since? Have you had occasion to meet or speak with them again?*

A number of the survivors have remained good friends over the more than thirty years since *Alive* was written—particularly Nando Parrado and Roberto Canessa. They almost always call when they come to London: once Roberto telephoned at ten in the evening to say that he had just hit town and was coming round for supper with his wife, Laura, two of his children, and three cousins! We also keep in touch with Pedro Algorta and his children have stayed on visits to Europe. Gustavo Zerbino has been to our house and this summer Coche Inciarte and his wife turned up with their son.

One of the most impressive aspects of Alive *is the restraint you show by not sensationalizing the aspects of the survivors' ordeal. You let the facts, as incredible as they are, speak for themselves. Did you find this aspect of the writing difficult —trying to balance writing a compelling book with respecting the truth of the events as they happened?*

When I sat down to write *Alive,* I had had no experience with nonfiction or reportage but I had the testimony of the survivors themselves still ringing in my ears. They had neither ▸

> "They had neither sensationalized nor sentimentalized their own experience and it seemed important for me to tell the reader what they had told me in the same 'matter-of-fact' manner."

sensationalized nor sentimentalized their own experience and it seemed important for me to tell the reader what they had told me in the same "matter-of-fact" manner and let the reader, after reliving their ordeal, decide how he or she would have reacted in the same circumstances. (When the survivors first read the manuscript of *Alive* some were dismayed that I had not told the reader what to think. I told them to have faith in my method and now they accept that it was right.)

Following up on the last question: Alive *is one of the pioneers of this genre of nonfiction. What do you think the responsibility of the writer is when retelling events such as those in* Alive? *Is there a code of ethics that should be followed, and do you ever question other writers' adherence to this code?*

There are certain technical skills involved in shaping material into a coherent narrative which apply to any work of nonfiction. When basing a book on the testimony of real people, the most important qualities are empathy, honesty, and a measure of self-restraint. The author must be a good listener: empathy means that he must have no preconceptions about those who are telling him their story, and he must not impose his own ideas or personality when it comes to retelling it in his book. In the case of *Alive,* it no doubt helped that both the author and the survivors shared the same religious beliefs—all were Roman Catholics—and the survivors were mostly likable young men. But even where there were differences between us—I hated rugby at school!—I did my best to remain dispassionate. After all, the reader wants to learn about the characters in the story, not the author.

In the introduction to Alive, *you mention that each survivor felt that you "wholly failed to understand him," but that you captured the character of the other survivors perfectly. Did this serve as a vindication for you—knowing that you must have done your job well if they invariably felt this way?*

This reaction of some of the individual survivors—that *he* was the only one whose personality had eluded me—taught me something about human nature in general—namely that one tends to form detached judgments about almost anyone except oneself! Vanity is part of the human condition and certainly writers are no exception!

Have the survivors shared any change in perception of how they were depicted in the book, looking back on it now?

I started talking to the survivors about their ordeal within six weeks or so after their return to Montevideo. Inevitably, they were still in a state of mixed shock and elation. As the months passed, there was a slow return to normality—the semi-mystical sense of closeness to God diminished as they were drawn back into everyday life. However, the experience has left an indelible mark on both their lives and their personalities. Nando Parrado, though he went on to become a successful businessman and television presenter, frequently speaks to gatherings around the world on what he learned from the experience. Coche Inciarte, too.

Clearly, for those whose role on the mountain as depicted in the book was less than heroic—Roy Harley, for example, or Pancho Delgado—the great success of the book, and later the appearance of the film, was a difficult experience—not just for them but for their children. But the weaknesses they may have shown in those extreme conditions (weaknesses which most of us may have shared) are not evident in their later lives. Pancho Delgado, so mistrusted on the mountain, is now employed by the survivors as their lawyer!

Alive *is such a compelling, heartbreaking, and inspiring story that just reading it has such a profound effect. You were intimately involved with these young men soon after their ordeal, and you lived closely with this story as you researched and wrote the book. How did knowing these men and being entrusted with their story change you? Do you still look back on the experience all of these years later as something that stands out in your life?*

I was thirty-one years old when I went out to Montevideo to meet the survivors and discuss the possibility of writing a book. I had led, hitherto, a privileged, protected, and somewhat self-indulgent life. Listening to the survivors' account of their ordeal was an exacting and disturbing experience. I learned from them to distinguish between the essential and inessential ingredients of human happiness—love of one's family, love of one's friends, and a surrender to the will of God. The words written by Arturo Nogueira in his letter to his *novia* have become a constant prayer. "Strength. Life is hard but it is worth living. Even suffering, courage."

Survivors Speak

HARPER PERENNIAL had the privilege of conducting an interview with two of the Andes survivors, Coche Inciarte and Alvaro Mangino. Find out what they have to say about their experience, more than thirty years later, in this exclusive interview.

It has been more than thirty years since your experience, and you have now lived a lifetime since the accident—what have you done since your rescue?

“We all dreamed of the same things, and that’s how we were able to endure everything we went through. What I wanted the most was to get out of there to be with my family and start my very own.”

Coche Inciarte (CI): Live! Live everything I thought I would never get a chance to live! That’s why I got married within eight months of our rescue, and had my first child just one year later. I started a family. That was what I missed the most during those seventy-two days . . . my family, my friends, the places I love. . . . We all dreamed of the same things, and that’s how we were able to endure everything we went through. What I wanted the most was to get out of there to be with my family and start my very own.

After the rescue, my life went on like anyone else’s. I married Soledad, who was my girlfriend back then, and now we have three wonderful children: Jose Luis, thirty, Maria Soledad, twenty-five, and Maria Eugenia, twenty-two. They are what I treasure most in the world, and the most wonderful gift of my entire life.

Professionally speaking, I became an agronomic engineer in 1973, and for thirty-five years I worked on a cattle farm owned by my family. I specialized in milk production as well as agriculture. Later, I became a member of the National Association of Milk Producers and in 1987, I was appointed the director of Conaprole, the largest dairy products company in my country, where I worked until 1997.

Now I'm retired, and since I have more time, I've decided to try myself at painting. That has been a dream of mine for many years, and up until now, I had never had a chance to make it come true. So for the past two and a half years, I have been painting and that makes me extremely happy. In some of these paintings I try to express what I felt when I was up there on the mountain and during other rural and urban experiences I have had. . . .

Thirty years later, I have been able to let go of the painful memories and concentrate on the positive experiences we had up there. I have started participating in Alvaro Mangino's motivational conferences, and this has taught me how to share my experiences and feelings with the others who were up there with me. It has been extremely helpful to talk about it with others, and we also feel that it helps the people who listen to us tell our stories. In the end, we feel that we learn a lot more from those who listen to us than what they learn from us.

Alvaro Mangino (AM): The truth is, it was very difficult to survive. I broke my left leg, and it was completely loose below my knee. But I think that from the first moment, I decided I wanted to live, and managed to do just that. I never stopped moving . . . dragging myself to an airplane seat every day during seventy-two days. . . . The whole time, I tried to hold on to the things that are dearest to me: my mother, my girlfriend, my faith, and the strength I needed to get through every single day I spent up there.

I married Margarita, who was my girlfriend at the time, and together we had four children: ▶

"I broke my left leg, and it was completely loose below my knee. But I think that from the first moment, I decided I wanted to live, and managed to do just that."

Survivors Speak *(continued)*

Daniela, twenty-nine, Federico, twenty-six, Margarita, twenty-one, and Felipe, eighteen. I struggled a lot—and continue to struggle—to have such a wonderful family. We lived outside Uruguay for some time, but now we're back in Montevideo.

After many years of not being able to talk about what happened, two years ago Coche and I decided to start talking to people about what helped us survive when we were on the mountain. It helps us a lot to be able to tell people how two young men managed to survive such an odyssey.

At the time Alive *was written, it had only been about a year since you were rescued, and it was still difficult to tell whether your health would permanently suffer from living in such extreme conditions for so long. Have there been any permanent health effects from this experience? Have the survivors as a group suffered any particular shared afflictions?*

CI: There have been no psychological effects because I truly believe that when we were up there, it was like being in therapy, all together, helping one another. Seventy-two days is a very long time to adapt to such a difficult environment and what we had to live through. But that's what's so wonderful about human beings, they are so adaptable. . . . Some of the survivors had physical consequences, like broken legs that healed while we were up on the mountain, and thus never healed properly. Others had to have eye surgery to not lose their eyesight because the bright reflection of the snow burned their retinas. But in the end, those are small things, and they managed to live their lives without troubles.

Once again, I have to say that for a long time the memories were so painful that I was unable to talk about it. But now, the bad memories have lost their power, and I can remember all the courage, bravery, love, and positive things we accomplished while we were up there and that's what I try to convey in the conferences. And that helps!

AM: I think that over the years, each and every one of us has gone through his or her own cathartic therapy. We never had to seek professional help, but each one of us reacted in a different way. In my case, for many years, I avoided talking about it in public, but now I have come to feel a lot better about sharing it with others.

We all have our own mountains, and it's important to remember that no matter how bad things are, one can always overcome them and more so, one must never forget that they can always be worse. It's important to value the small things in life.

Have you ever met with any survivors of catastrophic events such as yours from around the world? And do you ever feel a kinship when you hear of people who have lived through similar exceptional ordeals?

AM: I have met people who have survived other airplane accidents, but none of them had to survive for such a long time in a situation as difficult as ours.

There might be people who have had much more difficult experiences . . . I can't even imagine how hard it must be to be a survivor of the World War II concentration camps, or from the war. . . . The only thing I know is that words cannot even begin to explain what we had to go through up there.

CI: No, I have never met anyone who has been through anything remotely similar, I don't think anything like this has ever happened anywhere else. By this, I don't mean to say that no one has ever been though anything worse. Atrocious wars take place throughout the world, men fighting men. . . . Even though in our case, we were only fighting against nature, it was tremendously long and difficult.

At the time of the accident you were quite young, and you've lived a lifetime since then. Has anything in your life come close to achieving the importance and weight of the Andes experience, or do you feel that your experience on the mountain was so monumental that no event since could ever come close to matching those seventy-two days? ▸

> "We all have our own mountains, and it's important to remember that no matter how bad things are, one can always overcome them and more so, one must never forget that they can always be worse. It's important to value the small things in life."

"Those were seventy-two tragic days, but there were also some fantastic days in which I felt extraordinary things that I have never had a chance to feel again. I will never—and never want to—forget those seventy-two days."

Survivors Speak ***(continued)***

CI: I think seventy-two days out of my fifty-two years of age is really nothing, but they were so intense. I have never been through anything that intense in my entire life, though I must say that the day my first child was born was probably the most important day of my life.

Those were seventy-two tragic days, but there were also some fantastic days in which I felt extraordinary things that I have never had a chance to feel again. I will never—and never want to—forget those seventy-two days. They were moments in which I felt so proud to be a human being and have a chance to share with others the miracle of life.

AM: With no doubt, those were the worst seventy-two days of my life. But I also think that is where I learned about human nature, the importance of teamwork, and that the love we had for one another was what saved us. . . . It was extremely intense.

I have had to face other difficult moments in my life, and happy moments as well, like the days my children were born. . . . But in the end, I think this experience made me a happier man.

In Alive, *it is mentioned that while waiting to be rescued, the group often talked about how you would want to write a book about your experience when you returned to civilization. The title you chose was* Maybe Tomorrow, *because the hope that maybe tomorrow salvation would come was what kept you going for all those weeks. When you agreed to collaborate on* Alive, *did you still want to use the title* Maybe Tomorrow? *Did you feel that* Alive *was a better title, or would you have preferred to use* Maybe Tomorrow?

AM: I think *Alive* is the best possible title. I'll never forget that when my father arrived in Chile with my mother and Margarita, my girlfriend, and saw my name on the list of survivors, he yelled "He's ALIVE, for God's sake!"

CI: *Alive* is the best title because the book celebrates life.

What rituals, if any, have you followed since the accident to remember those who died? Do you still meet as a group each year, or play a rugby game, or hold a special mass each year to honor and remember the deceased?

CI: Today, as I write this it's October 13, 2004. Thirty-two years have passed since the accident, and I have not yet forgotten my friends that died on the mountain. In a few days we will celebrate a mass in their memory. On October 29 we remember the avalanche and on December 12 we have a great party celebrating the rescue. But it's difficult for all sixteen of us to get together, there is always someone missing. Overall, we do manage to see one another quite often and I stay in touch with some more than others.

I have already been up there six times to visit the resting place of my friends. Last March, I went with my family. It is always a very emotional experience, and I try to imagine what it would have been like for the young man I was back then, if he had known that one day he would come back with his family.

AM: We always remember those who did not make it back with us. Every year, we have a commemorative mass on October 13, and then on October 29, the day of the avalanche. I have been back there several times, and it's a place that gives me so much energy. I have gone back there with my family, my nephew who is the Capitan's godson and now my adoptive godson. Together we cried out of pain and joy, paying our respects to those heroic human beings.

You entrusted Piers Paul Read to tell your amazing story, which must have lead to a unique bond with the writer. Are you still in contact with him?

CI: At the beginning, I felt like Mr. Read was invading my privacy. At the time, I barely spoke to him, and didn't want to tell him much about what had happened. But after I read *Alive* I realized he had done a truly wonderful job in telling our story. This past June I had lunch with him at

Survivors Speak ***(continued)***

his house in Holland Park, London. I went there with my wife and son, and that day I was able to tell him so much more than thirty years ago. It was a real pleasure to see him again.

Read wrote this story and shared it with the world. In the thousands of e-mails we receive, I can tell that it has been very helpful for people who have read it, and that makes me very happy!

" Read wrote this story and shared it with the world. In the thousands of e-mails we receive, I can tell that it has been very helpful for people who have read it, and that makes me very happy! "

I find that it is very important to communicate the values and fundamental principles of human beings and life, and *Alive* does it so marvelously well. This exactly, is what Alvaro and I try to tell people at our conferences. And this is something that helps them, and helps us as well. Our friends who died on the mountain can't do it, but we can.

Will we ever be able to answer the eternal question of why me and what for?

AM: I didn't say much to Piers when he interviewed me. At the time, I felt that it was too personal. In the past years, I have not been in touch with Piers, but I do believe he did a fantastic job in documenting our story.

Please feel free to include anything else about your life before, during, or since your Andes experience that you would like readers to know.

AM: What I'd like to add, is that today, I feel great when Coche and I go out there and share our experience with others. I feel good because I can tell that people are truly thankful for the hope and the faith we give them. . . . Ultimately, maybe that's why this had to happen to us.

Beyond the Book

Learn more about the plane from Alive

FAIRCHILD FH-227 SPECS

First flight: 1966

Series: A, B, C, D, E

Production total: 78 built

Propulsion: 2 prop engines

Production ended: 1968

Maximum number of passengers: 56

Fatalities: There have been a total of 393 reported fatalities as the result of 25 incidents

FH-227 information courtesy of the Aviation Safety Network (http://aviation-safety.net)

Courtesy of Werner Fischdick

A plane very similar to the one that crashed in the Andes

Alive in Hollywood

In 1991, Oscar-nominated director Frank Marshall was trying to choose his next project. Jeffrey Katzenberg, then the chairman of Disney Studios, was trying to get Marshall on board to direct the film adaptation of *Alive,* based on the book of the same name. As he and his wife were out on a Sunday afternoon drive, out of the blue, a red pickup truck swerved in front of them, adorned with a bumper sticker that read "Rugby Players Eat Their Dead." Providence had spoken, and right then and there, they called Katzenberg to accept the project (*New York Times,* January 10, 1993, Mimi Avins).

Like Peirs Paul Read's book, the film version of *Alive* is a suspenseful, tragic, and inspiring tale of human endurance and fortitude. The film opens with the iconic Oscar-nominated actor John Malkovich, playing an older Carlitos Páez, recounting the events of the ill-fated plane crash that had changed his life forever. The viewer is then transported to the plane, where the rugby team and their friends and relatives are enjoying a carefree flight, unaware of the tragedy ahead. The film goes on to closely follow Read's book, including the cannibalism that sustains the survivors. Oscar-nominee Ethan Hawke plays Nando Parrado, the hero who leads the expedition across the Andes to find help, in a stirring performance that captures Parrado's indefatigable spirit and determination. Filmed in the Canadian Rockies, on an extremely realistic set (Nando Parrado was a technical advisor on the shoot), the film brings to life the incredible odds against these brave men's survival. The cliffs of ice and blinding white snow, the blistering cold wind, the confinement of the tiny fuselage that was their home for more than seventy days—all of these elements are painstakingly re-created, and the viewer cannot help but wonder how these men were able to survive their ordeal.

Watch *Alive* the movie, to see this compelling story brought to life in painstaking and breathtaking detail.

If You Liked *Alive*, You'll Love

Touching the Void

The *New York Times* Bestseller

Readers of *Alive* come away with an awe-inspiring feeling that the human spirit can triumph over impossible odds. From the compassion shown by those who helped care for their injured teammates for seventy-two days in extreme conditions, to the superhuman trip that Nando Parrado and Roberto Canessa undertook to find help in the valley, each page is a new lesson in resilience and love.

In *Touching the Void*, Joe Simpson and his climbing partner, Simon Yates, had just reached the top of a 21,000-foot peak in the Andes, when disaster struck. Simpson plunged off the vertical face of an ice ledge, breaking his leg. In the hours that followed, darkness fell and a blizzard raged as Yates tried to lower his friend to safety. Finally, Yates was forced to cut the rope, moments before he would have been pulled to his own death.

The next three days were an impossibly grueling ordeal for both men. Yates, certain that Simpson was dead, returned to base camp consumed with grief and guilt over abandoning him. Miraculously, Simpson had survived the fall, but crippled, starving, and severely frostbitten was trapped in a deep crevasse. Summoning vast reserves of physical and spiritual strength, Simpson crawled over the cliffs and canyons of the Andes, reaching base camp hours before Yates had planned to leave.

Much like the story of *Alive*, the story of how both men overcame the torments of those harrowing days is an epic tale of fear, suffering, and survival, and a poignant testament to unshakable courage and friendship.

"Told with a lyrical quality and stunning immediacy, *Touching the Void* transcends its genre and becomes accessible to readers who have never had any desire to climb."

—*New York Newsday*

"A document of psychological, even philosophical, witness to the rarest compulsion."

—*The Sunday Times* (London)

The Web Detective

Visit These Great Sites
to Further Explore the Themes of *Alive*

http://members.aol.com/porkinsr6/alive.html
A wonderful site that includes info on the crash, the movie, the plane, and updates on the survivors since their rescue

http://www.imdb.com/title/tt0106246/
The Internet Movie Database page for the film version of Alive

http://www.blueplanetbiomes.org/andes.htm
Information about the Andes mountains, including climate, vegetation, animals, and geography

http://www.cia.gov/cia/publications/factbook/geos/uy.html
Detailed information about Uruguay

http://www.pbs.org/wgbh/amex/donner/
A wonderful site about the Donner Party, another group of struggling survivors who chose to eat their dead in order to live

http://aviation-safety.net/index.shtml
A comprehensive site featuring information on aircraft safety

http://www.apple.com/trailers/independent/touching_the_void.html
Watch the movie trailer for the gripping survival film Touching the Void